Lecture Notes in Computer Science

Edited by G. Goos, J. Hartmanis, and J. van Leeuwen

Berlin
Heidelberg
New York
Barcelona
Hong Kong
London
Milan
Paris
Singapore
Tokyo

Cyril Fonlupt   Jin-Kao Hao   Evelyne Lutton
Edmund Ronald   Marc Schoenauer (Eds.)

Artificial Evolution

4th European Conference, AE '99
Dunkerque, France, November 3-5, 1999
Selected Papers

Series Editors

Gerhard Goos, Karlsruhe University, Germany
Juris Hartmanis, Cornell University, NY, USA
Jan van Leeuwen, Utrecht University, The Netherlands

Volume Editors

Cyril Fonlupt
LIL - Université du Littoral - Côte d'Opale
BP 719, 62228 Calais Cedex, France
E-mail: fonlupt@lil.univ-littoral.fr

Jin-Kao Hao
LERIA - Université d'Angers
2 Boulevard Lavoisier, 49045 Angers Cedex 01, France
E-mail: Jin-Kao.Hao@univ-angers.fr

Evelyne Lutton
INRIA Rocquencourt, Projet FRACTALES
Domaine de Voluceau, BP 105, 78153 Le Chesnay Cedex, France
E-mail: evelyne.lutton@inria.fr

Edmund Ronald
Marc Schoenauer
Ecole Polytechnique - Centre de Mathématiques Appliquées
91128 Palaiseau Cedex, France
E-mail: {eronald/Marc.Schoenauer}@cmapx.polytechnique.fr

Cataloging-in-Publication Data applied for

Die Deutsche Bibliothek - CIP-Einheitsaufnahme

Artificial evolution : 4th European conference ; selected papers / AE
'99, Dunkerque, France, November 2 - 5, 1999. Cyril Fonlupt ... (ed.).
- Berlin ; Heidelberg ; New York ; Barcelona ; Hong Kong ; London ;
Milan ; Paris ; Singapore ; Tokyo : Springer, 2000
    (Lecture notes in computer science ; Vol. 1829)
    ISBN 3-540-67846-8
CR Subject Classification (1998): F.1, F.2.2, I.2.6, I.5.1, G.1.6, J.3

ISSN 0302-9743
ISBN 3-540-67846-8 Springer-Verlag Berlin Heidelberg New York

Springer is a company in the BertelsmannSpringer publishing group.
© Springer-Verlag Berlin Heidelberg 2000
Printed in Germany

Typesetting: Camera-ready by author, data conversion by PTP-Berlin, Stefan Sossna
Printed on acid-free paper   SPIN: 10721187   06/3142   5 4 3 2 1 0

Preface

The Artificial Evolution conference was originally conceived as a forum for the French-speaking Evolutionary Computation community, but has of late been acquiring an European audience, with several papers from Germany, Austria, Italy, Spain... However, AE remains as intended a small and friendly gathering, which will continue to be held every two years.

Previous AE meets were held in Toulouse, Brest, and Nîmes. This year, the hosting was done by the LIL (Laboratoire d'Informatique du Littoral) in the not-so-cold city of Dunkerque.

The invited talk on "Fitness Landscapes and Evolutionary Algorithms" was delivered by Colin Reeves of Coventry University

This volume contains a selection of the papers presented at the conference. Twenty-seven papers were presented orally at the conference, selected from over 40 papers refereed by the program committee. After the conference, each presentation was reviewed and 20 papers were retained and revised for publication in this volume.

The papers in this volume have been grouped into the following five sections which more or less reflect the organization of the oral presentations.

1. **Invited Paper**: C. Reeves brightly describes the state of the art in Fitness Landscapes.
2. **Genetic Operators and Theoretical Models**: Devising new genetic operators and understanding the behavior of genetic operators is a popular research topic in evolutionary computation. Jens Gottlieb presents new initialization routines and several repair and optimization methods for the multidimensional knapsack problem. Jens Gottlieb and Günther Raidl characterize locality by analyzing the relation between genotypes and phenotypes. Mike Rosenman provides a solution to the problem of adaptation in case-based design. Aniko Ekart introduces a special mutation operator in order to moderate code growth in genetic programming.

   In the only theoretical contribution, Anton Eremeev proposes a mathematical model of a simplified GA and obtains upper and lower bounds on the expected proportion of individuals above a given threshold.
3. **Applications**: This section demonstrates the successful applicability of EAs in a broad range of problems. Monmarché *et al.* generate style sheets for web sites based on artistic preferences with the help of a GA; Alain Ratle maximizes the absorbing properties of a composite material using an EA to find out the optimal distribution of two or more types of sound absorbing material elements; Laurence Moreau-Giraud and Pascal Lafon present their hybrid evolution strategy for mixed discrete continuous constrained problems; Anne Spalanzani proposes an evolutionary solution for speech recognition problems; Jean Louchet proposes a new swarm-based algorithm

for fast 3D image analysis; Yu Li and Youcef Bouchebaba solve the optimal communication spanning tree problem by means of evolutionary techniques.

4. **Agents - Cooperation:** Alternative evolutionary paradigms are introduced in this section. Philippe Mathieu *et al.* present a classification of ecological evolution by studying the classical iterated prisoner's dilemma. A.J. Bagnall and G.D. Smith describe an autonomous agent model of the UK market in electricity. Samuel Delepoulle *et al.* study the evolution of social behaviors within a behavioral framework, and David Griffiths and Anargyros Serafopoulos present some animated sequences of simulated agent colonies.

5. **Heuristics - Outlooks**: This section collects studies reflecting the general experience of the authors regarding evolutionary computation. Olivier Roux *et al.* use the ant colony paradigm to solve the quadratic assignment problem. Meriema Belaidouni and Jin-Kao Hao use the fitness landscape paradigm in order to study the difficulty of a combinatorial problem. Philippe Collard *et al.* introduce the notion of synthetic neutrality, an original paradigm for describing a problem difficulty. Sana Ben Hamida *et al.* compare the plain parametric approach to the GP parse-tree representation for the design of a 2-dimensional profile of an optical lens in order to control focal-plane irradiance of some laser bean. Denis Robilliard and Cyril Fonlupt use a repellent-attractor strategy to guide evolutionary computation.

At this point, we would like to mention Denis Robilliard, Philippe Preux, Olivier Roux, and Eric Ramat at the LIL in Calais, and thank them for their invaluable assistance with the nuts and bolts of organizing AE'99. Finally, we would like to thank the AE'99 program committee members for the service they rendered to the community by ensuring the high scientific content of the papers presented. The names of these very busy people, who still found time or made time to do the refereeing, are listed on the following pages.

April 2000

Cyril Fonlupt,
Jin-Kao Hao,
Evelyne Lutton,
Edmund Ronald,
and Marc Schoenauer.

Artificial Evolution 99 - EA'99
November 3-5, 1999
LIL, Université du Littoral, Dunkerque, France

AE'99 is the fourth conference on Evolutionary Computation organized in France. Following EA'94 in Toulouse, EA'95 in Brest, and EA'97 in Nîmes, the conference was held in Dunkerque.

AE'99 was hosted by the LIL, Laboratoire d'Informatique du Littoral.

Organizing Committee

Cyril Fonlupt (LIL Calais) - Jin-Kao Hao (Université d'Angers)
Evelyne Lutton (INRIA Rocquencourt) - Edmund Ronald (Ecole Polytechnique Palaiseau)
Marc Schoenauer (Ecole Polytechnique Palaiseau)

Program Committee

Jean-Marc Alliot (ENAC Toulouse) - Thomas Bäck (Informatik Centrum Dortmund - Leiden University)
Pierre Bessiere (LIFIA Grenoble) - Paul Bourgine (CREA Palaiseau)
Bertrand Braunschweig (IFP Rueil Malmaison) - Philippe Collard (Université de Nice)
David Corne (Reading University) - Marco Dorigo (ULB Bruxelles)
Reinhardt Euler (UBO Brest) - David Fogel (Natural Selection Inc. La Jolla)
Hajime Kita (Kyoto University) - Jean Louchet (ENSTA Paris)
Bernard Manderick (VUB Bruxelles) - Zbigniew Michalewicz (UNCC Charlotte)
Olivier Michel (EPFL Lausanne) - Francesco Mondada (EPFL Lausanne)
Philippe Preux (Université du Littoral) - Nick Radcliffe (Quadstone Edinburgh)
Denis Robilliard (Université du Littoral) - Günter Rudolph (Dortmund University)
Michele Sebag (LMS Palaiseau - LRI Orsay) - Moshe Sipper (EPFL Lausanne)
Mohammed Slimane (E3I, Université de Tours) - El-Ghazali Talbi (Université de Lille)
Gilles Venturini (E3I, Université de Tours)

Highlights

An invited talk by Colin Reeves (Coventry University)

27 paper presentations

Sponsoring Institutions

Université du Littoral - Côte d'Opale
Délégation générale à l'armement (DGA)
Ministère de l'Education Nationale de la Recherche et de la Technologie (MENRT)
Conseil régional du Nord - Pas-de-Calais
Conseil général du Nord

Table of Contents

Invited Paper

Genetic Operators and Theoretical Models

Applications

Agents - Cooperation

Heuristics - Outlooks

Fitness Landscapes and Evolutionary Algorithms

Colin R. Reeves

School of Mathematical and Information Sciences
Coventry University
Priory Street, Coventry CV1 5FB
UK
C.Reeves@coventry.ac.uk,
http://www.mis.coventry.ac.uk/~colinr/

Abstract. Evolutionary algorithms (EAs) have been increasingly, and successfully, applied to combinatorial optimization problems. However, EAs are relatively complicated algorithms (compared to local search, for example) and it is not always clear to what extent their behaviour can be explained by the particular set of strategies and parameters used.

One of the most commonly-used metaphors to describe the process of simple methods such as local search is that of a 'fitness landscape', but even in this case, describing what we mean by such a term is not as easy as might be assumed.

In this paper, we first present some intuitive ideas and mathematical definitions of what is meant by a landscape and its properties, and review some of the theoretical and experimental work that has been carried out over the past 6 years. We then consider how the concepts associated with a landscape can be extended to search by means of evolutionary algorithms, and connect this with previous work on epistasis variance measurement.

The example of the landscapes of the *Onemax* function will be considered in some detail, and finally, some conclusions will be drawn on how knowledge of typical landscape properties can be used to improve the efficiency and effectiveness of heuristic search techniques.

1 Introduction

The metaphor of a *landscape* is commonly found in descriptions of the application of heuristic methods for solving a combinatorial optimization problem (COP). We can define such problems as follows: we have a discrete search space $\mathcal{X}$, and a function

$$f : \mathcal{X} \mapsto \mathbb{R}.$$

The general problem is to find

$$x^* = \max_{x \in \mathcal{X}} f.$$

where x is a vector of *decision variables* and f is the *objective function*. (Of course, minimization can also be the aim, but the modifications are always obvious). In the field of evolutionary algorithms, such as genetic algorithms (GAs),

C. Fonlupt et al. (Eds.): AE'99, LNCS 1829, pp. 3–20, 2000.

the function f is often called the *fitness*, and the associated landscape is a *fitness landscape*. The vector $\mathbf{x}^*$ is a global optimum: that vector which is the 'fittest' of all. (In some problems, there may be several global optima—different vectors of equal fitness.)

With the idea of a fitness landscape comes the idea that there are also many local optima or false peaks, in which a search algorithm may become trapped without finding the global optimum. In continuous optimization, notions of continuity and concepts associated with the differential calculus enable us to characterize quite precisely what we mean by a landscape, and to define the idea of an optimum. It is also convenient that our own experiences of hill-climbing in a 3-dimensional world gives us analogies to ridges, valleys, basins, watersheds etc that help us to build an intuitive picture of what is needed for a successful search, even though the search spaces that are of interest often have dimensions many orders of magnitude higher than 3.

However, in the continuous case, the landscape is determined only by the fitness function, and the ingenuity needed to find a global optimum consists in trying to match a technique to this single landscape. There is a major difference when we come to discrete optimization. Indeed, we really should not even use the term 'landscape' unless we can define the topological relationships of the points in the search space $\mathcal{X}$. Unlike the continuous case, we have some freedom to specify these relationships, and in fact, that is precisely what we do when we decide to use a particular technique.

1.1   An Example

In practice, one of the most commonly used search methods for COPs is *neighbourhood search* (NS). This idea is at the root of modern 'metaheuristics' such as simulated annealing (SA) and tabu search (TS)—as well as being much more involved in the methodology of GAs than is sometimes realized.

A *neighbourhood structure* is generated by using an operator that transforms a given vector $\boldsymbol{x}$ into a new vector $\boldsymbol{x}'$. For example, if the solution is represented by a binary vector (as is often so for GAs, for instance), a simple neighbourhood might consist of all vectors obtainable by 'flipping' one of the bits. The 'bit flip' (BF) neighbours of (00000), for example, would be

$$\{(10000), (01000), (00100), (00010), (00001)\}.$$

Consider the problem of maximizing a simple function

$$f(z) = z^3 - 60z^2 + 900z + 100$$

where the solution z is required to be an integer in the range $[0, 31]$. Regarding z as a continuous variable, we have a smooth unimodal function with a single maximum at $z = 10$—as is easily found by calculus—and since the solution is already an integer, this is undoubtedly the most efficient way of solving the problem.

However, suppose we chose instead to represent z by a binary vector $\boldsymbol{x}$ of length 5. By decoding this binary vector as an integer it is possible to evaluate

f, and we could then use NS, for example, to search over the binary hypercube for the global optimum using some form of hill-climbing strategy.

Table 1. Local optima and basins of attraction for steepest ascent with the BF operator in the case of a simple cubic function. The bracketed figures are the fitnesses of each local optimum.

Local optimum	0 1 0 1 0	0 1 1 0 0	0 0 1 1 1	1 0 0 0 0
	(4100)	(3988)	(3803)	(3236)
Basin	0 0 0 0 0	0 0 1 0 0	0 0 1 1 0	1 0 0 0 0
	0 0 0 0 1	0 1 1 0 0	0 0 1 1 1	1 0 0 0 1
	0 0 0 1 0	1 1 1 0 0	1 0 1 1 0	1 0 0 1 0
	0 0 0 1 1		1 0 1 1 1	1 0 0 1 1
	0 0 1 0 1			1 0 1 0 0
	0 1 0 0 0			
	0 1 0 0 1			
	0 1 0 1 0			
	0 1 0 1 1			
	0 1 1 0 1			
	0 1 1 1 0			
	0 1 1 1 1			
	1 0 1 0 1			
	1 1 0 0 0			
	1 1 0 0 1			
	1 1 0 1 0			
	1 1 0 1 1			
	1 1 1 0 1			
	1 1 1 1 0			
	1 1 1 1 1			

This discrete optimization problem turns out to have 4 optima (3 of them local) when the BF operator is used. If a 'steepest ascent' strategy is used (i.e., the *best* neighbour of a given vector is identified before a move is made) the local optima and their basins of attraction are as shown in table 1. On the other hand, if a 'next ascent' strategy is used (where the next change which leads uphill is accepted without ascertaining if a still better one exists), the basins of attraction are as shown in table 2.

In fact, there are even more complications: in table 2, the order of searching the components of the vector is 'forward' (left-to-right). If the search is made in the reverse direction (right-to-left) the basins of attraction are different, as shown in table 3.

Thus, by using BF with this binary representation, we have created local optima that did not exist in the integer version of the problem. Further, although the optima are still the same the chances of reaching a particular optimum can be seriously affected by a change in hill-climbing strategy.

Table 2. Local optima and basins of attraction for next ascent (forward search) using the BF operator.

Local optimum	0 1 0 1 0 (4100)	0 1 1 0 0 (3988)	0 0 1 1 1 (3803)	1 0 0 0 0 (3236)
Basin	0 0 1 0 1	0 0 1 0 0	0 0 1 1 1	0 0 0 0 0
	0 0 1 1 0	0 1 0 0 0	0 1 1 1 1	0 0 0 0 1
	0 1 0 0 1	0 1 1 0 0	1 0 1 1 1	0 0 0 1 0
	0 1 0 1 0	1 0 1 0 0	1 1 1 1 1	0 0 0 1 1
	0 1 0 1 1	1 1 0 0 0		1 0 0 0 0
	0 1 1 0 1	1 1 1 0 0		1 0 0 0 1
	0 1 1 1 0			1 0 0 1 0
	1 0 1 0 1			1 0 0 1 1
	1 0 1 1 0			
	1 1 0 0 1			
	1 1 0 1 0			
	1 1 0 1 1			
	1 1 1 0 1			
	1 1 1 1 0			

Table 3. Local optima and basins of attraction for next ascent (reverse search) using the BF operator.

Local optimum	0 1 0 1 0 (4100)	0 1 1 0 0 (3988)	0 0 1 1 1 (3803)	1 0 0 0 0 (3236)
Basin	0 1 0 0 0	0 1 1 0 0	0 0 0 0 0	1 0 0 0 0
	0 1 0 0 1	0 1 1 0 1	0 0 0 0 1	1 0 0 0 1
	0 1 0 1 0	0 1 1 1 0	0 0 0 1 0	1 0 0 1 0
	0 1 0 1 1	0 1 1 1 1	0 0 0 1 1	1 0 0 1 1
			0 0 1 0 0	1 0 1 0 0
			0 0 1 0 1	1 0 1 0 1
			0 0 1 1 0	1 0 1 1 0
			0 0 1 1 1	1 0 1 1 1
				1 1 0 0 0
				1 1 0 0 1
				1 1 0 1 0
				1 1 0 1 1
				1 1 1 0 0
				1 1 1 0 1
				1 1 1 1 0
				1 1 1 1 1

However, the bit flip operator is not the only mechanism for generating neighbours. An alternative neighbourhood could be defined as follows: for $k = 1, \ldots, 5$, flip bits $\{k, \ldots, 5\}$. Thus, the neighbours of (0 0 0 0 0), for example, would now be

$$\{(11111), (01111), (00111), (00011), (00001)\}.$$

We shall call this the 'CX operator', and it creates a very different landscape. In fact, there is now only a single global optimum (01010); *every* vector is in its basin of attraction. This illustrates the point that it is not merely the choice of a binary representation that generates the landscape—the search operator needs to be specified as well.

Incidentally, there are two interesting facts about the CX operator. Firstly, it is closely related to the one-point crossover operator frequently used in genetic algorithms. (For that reason, it has been named [1] the complementary crossover or CX operator). Secondly, if the 32 vectors in the search space are re-coded using a *Gray* code, it is easy to show that the neighbours of a point in Gray-coded space under BF are identical to those in the original binary-coded space under CX. This is an example of an *isomorphism* of landscapes.

2 Mathematical Characterization

We can define a landscape Λ for the function f as a triple $\Lambda = (\mathcal{X}, f, d)$ where d denotes a distance measure $d : \mathcal{X} \times \mathcal{X} \to \mathbb{R}^+ \cup \{\infty\}$ for which is required that

$$d(s, t) \geq 0; \; d(s, t) = 0 \Leftrightarrow s = t; \; d(s, u) \leq d(s, t) + d(t, u); \} \quad \forall s, t, u \in \mathcal{X}.$$

Note that we do not need to specify the representation explicitly (for example, binary or Gray code), since this is assumed to be implied in the description of $\mathcal{X}$ (or alternatively, in that of f). We have also decided, for the sake of simplicity, to ignore questions of search strategy and other matters in the definition of a landscape, unlike the more comprehensive definition of Jones [2], for example.

This definition says nothing about how the distance measure arises. In fact, for many cases a 'canonical' distance measure can be defined. Often, this is symmetric, i.e. $d(s, t) = d(t, s) \; \forall s, t \in \mathcal{X}$, so that d also defines a *metric* on $\mathcal{X}$. This is clearly a nice property, although it is not essential.

2.1 Neighbourhood Structure

What we have called a canonical distance measure is typically related to the neighbourhood structure. Every solution $x \in \mathcal{X}$ has an associated set of *neighbours*, $N(x) \subset \mathcal{X}$, called the neighbourhood of x. Each solution $x' \in N(x)$ can be reached directly from x by an operation called a *move*. Many different types of move are possible in any particular case, and we can view a move as being generated by the applying an operator ω to a vector s in order to transform it into a vector t. The canonical distance measure d_ω is that induced by ω whereby

$$t \in N(s) \Leftrightarrow d_\omega(s, t) = 1.$$

The distance between non-neighbours is defined as the length of the shortest path between them (if one exists).

For example, if $\mathcal{X}$ is the binary hypercube $\mathbb{Z}_2^l$, the bit flip (BF) operator can be defined as

$$\phi(k) : \mathbb{Z}_2^l \to \mathbb{Z}_2^l \quad \begin{cases} z_k \mapsto 1 - z_k \\ z_i \mapsto z_i \end{cases} \quad \text{if } i \neq k$$

where z is a binary vector of length l. It is clear that the distance metric induced by ϕ is the well-known Hamming distance. Thus we could describe this landscape as a Hamming landscape (with reference to its distance measure), or as the BF landscape (with reference to the operator). Similarly, we can define the CX operator as

$$\gamma(k) : \mathbb{Z}_2^l \to \mathbb{Z}_2^l \quad \begin{cases} z_i \mapsto 1 - z_i \text{ for } i \geq k \\ z_i \mapsto z_i \quad \text{otherwise} \end{cases}$$

2.2   Local Optima

We can now give a formal statement of a fundamental property of fitness landscapes: for a landscape $\Lambda = (\mathcal{X}, f, d)$, a vector $s \in \mathcal{X}$ is *locally optimal* if

$$f(s) > f(t) \ \forall \ t \in N(s).$$

Landscapes that have only one local (and thus also global) optimum are commonly called *unimodal*, while landscapes with more than one local optimum are said to be *multimodal*.

The number of local optima in a landscape clearly has some bearing on the difficulty of finding the global optimum. However, it is not the only indicator: the size of the basins of attraction of the various optima is also an important influence.

2.3   Graph Representation

Neighbourhood structures are clearly just another way of defining a graph Γ, which can be described by its $(n \times n)$ *adjacency matrix* A. The elements of A are given by $a_{ij} = 1$ if the indices i and j represent neighbouring vectors, and $a_{ij} = 0$ otherwise. For example, the graph induced by the bit flip ϕ on binary vectors of length 3 has adjacency matrix

$$A_\phi = \begin{bmatrix} 0 & 1 & 1 & 0 & 1 & 0 & 0 & 0 \\ 1 & 0 & 0 & 1 & 0 & 1 & 0 & 0 \\ 1 & 0 & 0 & 1 & 0 & 0 & 1 & 0 \\ 0 & 1 & 1 & 0 & 0 & 0 & 0 & 1 \\ 1 & 0 & 0 & 0 & 0 & 1 & 1 & 0 \\ 0 & 1 & 0 & 0 & 1 & 0 & 0 & 1 \\ 0 & 0 & 1 & 0 & 1 & 0 & 0 & 1 \\ 0 & 0 & 0 & 1 & 0 & 1 & 1 & 0 \end{bmatrix}$$

where the vectors are indexed in the usual binary-coded integer order (i.e., $(000), (001)$ etc). By way of contrast, the adjacency matrix for the CX operator is

$$A_\gamma = \begin{bmatrix} 0 & 1 & 0 & 1 & 0 & 0 & 0 & 1 \\ 1 & 0 & 1 & 0 & 0 & 0 & 1 & 0 \\ 0 & 1 & 0 & 1 & 0 & 1 & 0 & 0 \\ 1 & 0 & 1 & 0 & 1 & 0 & 0 & 0 \\ 0 & 0 & 0 & 1 & 0 & 1 & 0 & 1 \\ 0 & 0 & 1 & 0 & 1 & 0 & 1 & 0 \\ 0 & 1 & 0 & 0 & 0 & 1 & 0 & 1 \\ 1 & 0 & 0 & 0 & 1 & 0 & 1 & 0 \end{bmatrix}$$

It is simply demonstrated that permuting the rows and columns so that they are in the order $0, 1, 3, 2, 6, 7, 5, 4$ reproduces the adjacency matrix A_ϕ—another way of demonstrating the isomorphism mentioned earlier. In other words,

$$P^{-1}A_\phi P = A_\gamma$$

where P is the associated permutation matrix of the binary-to-Gray transformation. It is also clear that the eigenvalues and eigenvectors are the same.

2.4   Laplacian Matrix

The *graph Laplacian* Δ is defined as

$$\Delta = A - D$$

where D is a diagonal matrix such that d_{ii} is the degree of vertex i. Usually, these matrices are vertex-regular and $d_{ii} = k \; \forall i$, so that

$$\Delta = A - kI.$$

This notion recalls that of a Laplacian operator in the continuous domain; the effect of this matrix, applied as an operator at the point s to the fitness function f is

$$\Delta f(s) = \sum_{t \in N(s)} (f(t) - f(s))$$

so it functions as a kind of differencing operator. In particular, $\Delta f(s)/|N(s)|$ is the average difference in fitness between the vector s and its neighbours. Grover [3] has shown that the landscapes of several COPs satisfy an equation of the form

$$\Delta f + \frac{Cf}{n} = 0$$

where C is a problem-specific constant and n is the size of the problem instance. From this it can be deduced that *all* local optima are better than the mean $(\bar{f})$ over all points on the landscape. Furthermore, it can also be shown that under mild conditions on the nature of the fitness function, the time taken by NS to find a local optimum in a maximization problem is $\mathcal{O}(n \log_2[f_{max}/\bar{f}])$ where f_{max} is the fitness of a global maximum. (A similar result can be obtained, *mutatis mutandis*, for minimization problems.)

2.5 Graph Eigensystem

In the usual way, we can define eigenvalues and eigenvectors of the matrices associated with a graph. The set of eigenvalues is called the *spectrum* of the graph. For an $n \times n$ matrix A the spectrum is

$$\left(\lambda_0 \; \lambda_1 \; \cdots \; \lambda_{n-1} \right)$$

where λ_i is the ith eigenvalue, ranked in (weakly) descending order. Similarly, the spectrum of the Laplacian is

$$\left(\mu_0 \; \mu_1 \; \cdots \; \mu_{n-1} \right)$$

where, again, μ_i is the ith eigenvalue, ranked this time in (weakly) ascending order. For a regular connected graph it can be shown that

$$\mu_i = k - \lambda_i \; \forall i.$$

Further, from the corresponding eigenvectors $\{\varphi_i\}$, f can be expanded as

$$f(s) = \sum_i a_i \varphi_i(s).$$

Stadler and Wagner [4] call this a 'Fourier expansion'. Usually, the eigenvalues are not simple, and this sum can be further partitioned into a sum

$$f(s) = \sum_p \beta_p \tilde{\varphi}_p(s)$$

over the distinct eigenvalues of $\mathbf{\Delta}$. The corresponding values

$$|\beta_p|^2 = \sum |a_k|^2$$

(where the sum is over the coefficients that correspond to the pth distinct eigenvalue) form the *amplitude spectrum*, which expresses the relative importance of different components of the landscape.

Ideally, such mathematical characterizations could be used to aid our understanding of the important features of a landscape, and so help us to exploit them in designing search strategies. But beyond Grover's rather general results above, it is possible to carry out further analytical studies only for small graphs, or graphs with a special structure. In the case of Hamming landscapes it is possible to find analytical results for the graph spectrum which show that the eigenvectors are thinly disguised versions of the familiar Walsh functions. For the case of recombinative operators the problem is rather more complicated, and necessitates the use of 'P-structures' [4]. The latter are essentially generalizations of graphs in which the mapping is from pairs of 'parents' $(\boldsymbol{x}, \boldsymbol{y})$ to the set of possible strings that can be generated by their recombination. However, it can be shown that for some 'recombination landscapes' (such as that arising from the use of uniform crossover) the eigenvectors are once more the Walsh functions. Whether this is also true in the case of 1- or 2-point crossover, for example, is not known, but Stadler and Wagner conjecture that it is. In view of the close relationship between the BF and CX landscapes as demonstrated above, it would not be surprising if this is a general phenomenon. However, to obtain these results, some assumptions have to be made—such as a uniform distribution of parents—that are unlikely to be true in a specific finite realization of a genetic search.

In the case of the BF landscape, the distinct eigenvalues correspond to sets of Walsh coefficients of different orders, and the amplitude spectrum is exactly the set of components of the 'epistasis variance' associated with other attempts to measure problem difficulty (see [5] for a review). For the cubic function of section 1.1 above, the components of variance for the different orders of Walsh coefficients can be shown to be $(0.387, 0.512, 0.101, 0, 0)$ respectively. It is clear that the interactions predominate and in this case indicate the relatively poor performance of the BF hill-climber.

Of course, the eigenvalues and eigenvectors are exactly the same for the CX landscape of this function, and the set of Walsh coefficients in the Fourier decomposition is also the same. However, the effect of the permutation inherent in the mapping from the BF landscape to the CX landscape is to re-label some of the vertices of the graph, and hence some of the Walsh coefficients. Thus some coefficients that previously referred to linear effects now refer to interactions, and vice-versa. Taking the cubic function as an example again, the components of variance or amplitude spectrum becomes $(0.771, 0.174, 0.044, 0.011, 0.000)$. We see that the linear effects now predominate, and this is consistent with the results we obtained from the CX hill-climber.

2.6   Recombination Landscapes

If we look at the 'recombination landscapes' derived from Stadler's and Wagner's P-structures [4], we find that once again the Walsh coefficients are obtained, but labelled in yet another way. The coefficients in the BF and CX landscapes are grouped according to the number of 1s in their binary- and Gray-coded index representations respectively. However, in a recombination landscape—such as that generated by 1-point crossover—it is the *separation* between the outermost 1-bits that defines the groupings. Table 4 shows the groupings for a 4-bit problem.

Several things can be seen from this table: firstly, the linear Walsh coefficients (and hence the linear component of epistasis variance) are the same in both the BF and the crossover landscapes. Secondly (as already explained), the coefficients in the CX landscape are simply a re-labelling of those in the BF landscape. Thirdly, the coefficients in the recombination landscape do not form a natural grouping in terms of interactions, and consequently the different components of variance for the recombination landscape do not have a simple interpretation as due to interactions of a particular order.

Table 4. Illustration of the different groupings of the Walsh coefficients associated with the BF, CX and recombination landscapes.

Index	binary coding	BF	CX	crossover	Index	binary coding	BF	CX	crossover
0	0000	0	0	0	8	1000	1	2	1
1	0001	1	1	1	9	1001	2	3	4
2	0010	1	2	1	10	1010	2	4	3
3	0011	2	1	2	11	1011	3	3	4
4	0100	1	2	1	12	1100	2	2	2
5	0101	2	3	3	13	1101	3	3	4
6	0110	2	2	2	14	1110	3	2	3
7	0111	3	1	3	15	1111	4	1	4

3   The *Onemax* Problem

As a second example, we can take the well-known *Onemax* problem: maximize

$$\sum_{i=1}^{l} x_i, \ x_i \in \{0,1\}, \tag{1}$$

This is obviously easy for a hill-climber in the BF landscape, and it is often claimed that this function is also 'GA-easy'. If we calculate the epistasis variances, both the BF and recombination landscapes have 100% in the linear component, so we would certainly expect it to be easy for a GA.

If we use a GA that relies exclusively on mutation, implemented by the BF operator, then it proceeds well at first, but slows down as it tries to fix the last few 1-bits. When crossover is brought into the picture, things are considerably more complicated. In practice, a GA that relies only on crossover does solve this problem, but compared to a hill-climber it is rather inefficient, as Figure 1 demonstrates. This shows the results of applying the following procedure: each algorithm was run from a random starting point until it converged; if the convergence point was not the global optimum, the process was repeated until the global optimum was obtained and the total number of function evaluations required was recorded. The whole procedure was repeated 30 times in order to reduce the inherent statistical variation.

For the BF landscape, no more than l function evaluations will ever be needed if we use a next ascent strategy. (In fact, this somewhat overstates the computational requirements, since the whole function does not actually need to be evaluated, and on average $l/2$ bits will be correct even before the search starts.) In the interests of clarity, the case of a BF hill-climber has been omitted from Figure 1.

In the limiting case, where 2 parents are maximally different, the recombination landscape is simply the CX landscape. For the case of *Onemax*, this can be shown (using an argument first put forward by Culberson [6]) to have a very large number of local optima—a number that increases exponentially with l. It

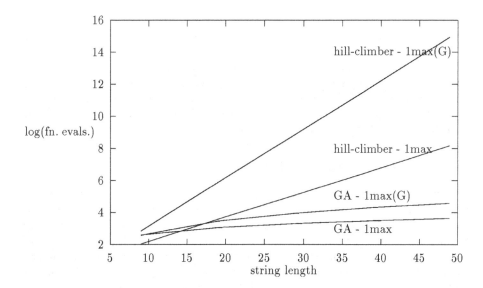

Fig. 1. Performance of various algorithms for *Onemax* and Gray-encoded *Onemax* functions. The vertical axis measures the number of function evaluations needed (averaged over 30 independent trials), and is on a logarithmic scale. The hill-climber used a next ascent strategy in the CX landscape (but note that the results for *Onemax* are identical to that of a hill-climber in the BF landscape of Gray-encoded *Onemax*. The GA was generational with binary tournament selection, one-point crossover (at a rate of 100%), no mutation and a population of 200—mutation was ignored because the focus here is on the recombination landscape.

is hardly surprising therefore, that a CX hill-climber needs an exponentially increasing amount of computation. This is also reflected by the fact that the linear component of epistasis variance tends to zero as l increases.

A comparison can also be made with the landscapes associated with the Gray-encoded *Onemax* function[1]. Because of the BF/CX isomorphism, a hill-climber in the BF landscape of the Gray-encoded *Onemax* will have the same performance as a hill-climber in the CX landscape does for the traditional binary-encoded *Onemax*—that is, an exponential increase in computational effort. What will happen for a hill-climber in the CX landscape of the Gray-encoded *Onemax* is not easy to predict, but in practice it seems to increase exponentially, although more slowly than in the case of binary-encoded *Onemax*. (A regression analysis suggests the rates of increase are roughly $(2.0)^l$ and $(1.4)^l$ respectively.)

Figure 1 also shows the results of applying a GA. Its performance seems to be considerably better than that of the CX hill-climbers in either landscape,

---

[1] By a Gray-encoded function we mean a function $g(x)$, say, defined by $f(G(x))$ where $f(\cdot)$ is the original function and $G(x)$ is the binary-to-Gray mapping.

if somewhat more than linear: a regression analysis suggests the computational requirements increase at a rate proportional to $l^{1.4}$ for *Onemax* and to $l^{2.7}$ for the Gray-encoded version.

We recall that the calculation of the linear epistasis variance component will not distinguish between the BF and recombination landscapes of *Onemax*. While it is clear that a BF hill-climber and a GA are not equally effective in solving it, given the difficulty faced by a CX hill-climber, the interesting question is: how exactly does the GA solve the *Onemax* problem at all?

4   How Recombination Solves *Onemax*

Our starting point is the CX landscape. This is clearly identical to the recombination landscape in the event that the parents are complementary pairs, but this is highly unlikely in practice! On the other hand, the landscape described by Stadler's and Wagner's P-structures essentially consists of all possible pairs of parents, uniformly distributed—again, not what we find in practice. Commonly, we start from a random initial population, so that in a particular realization of a GA not all pairings can be generated. For example, in the first generation the parents are on average $l/2$ bits different, and for large l the chance of a significant deviation from this number is very small. So the P-structure approach is interesting, but only in providing a general picture; what happens at a detailed level in a particular realization is what we are interested in.

4.1   Recombination's Dual Effect

The impact of recombination on the CX landscape is twofold. While the CX landscape is *connected* (i.e. the distance $d_\gamma(\boldsymbol{x}, \boldsymbol{y}) < \infty \; \forall \boldsymbol{x}, \boldsymbol{y} \in \mathbb{Z}_2^l$), the 1-point crossover landscape is composed of disconnected components defined by the vector $\boldsymbol{d} = \boldsymbol{x} \oplus \boldsymbol{y}$ (where $\oplus$ is addition modulo 2—the 'exclusive or' operator). In essence, recombination restricts the search to one of these components. We have elsewhere [1] called this the primary effect of recombination. The size of the components is determined by the Hamming distance between the parents. Since the average Hamming distance between parents shrinks during search—an effect that has often been observed in genetic algorithms—the search becomes restricted to smaller and smaller components. As conventional GA terminology would put it, the search proceeds from lower order to higher order schemata.

But that is not all that recombination can do: it also effects a particular type of linear transformation of the CX landscape in such a way that neighbourhoods (and therefore potentially local optima) are not preserved. In particular, its neighbourhood in the 1-point crossover landscape is, in general, not a subset of its neighbourhood in the CX landscape, as is illustrated in Figure 2. (This is not so if we apply a similar linear transformation to the BF landscape.)

This is the secondary effect of recombination, which is important for understanding what happens to local optima. We argue that by means of this secondary effect, the search can escape from local optima that are present in the underlying CX landscape by modifying the landscape as it proceeds. In [1] we described some experiments to verify that this occurs.

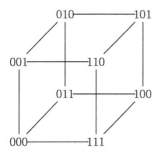
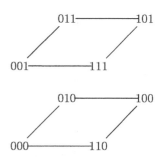

Fig. 2. Two effects of recombination: while the neighbourhood graph of the CX landscape (left) is connected that of the 1-point crossover landscape (right) consists of two subgraphs. Moreover, the vertex set of the latter is a permutation of the former, obtained by a linear transformation (secondary effect). In this case, $d = (1, 1, 0)^T$, e.g. the effect of crossing the binary strings (010) and (100).

4.2   The Importance of Search Strategy

However, the search strategy also has an important influence. After all, we could use *any* vector d to accomplish the primary effect of restricting the search space to a subspace. Why should pairwise mating of population members be an appropriate way of proceeding? The following experiments help to demonstrate the importance of this rule.

In Jones's 'headless chicken' experiments [7], only one parent was taken from the current population, while the other was selected from the entire search space, so that d could be almost any vector in $\mathbb{Z}_2^l$, rather than being restricted to the subset implied by the current population.

We repeated the 'headless chicken' experiments for both *Onemax* and its Gray-encoded version. Apart from these modifications in the parent selection scheme, the GA used in the following experiments was the same as before. Figure 3 summarizes the results.

For both cases here the GA's performance is clearly inferior when the range of parent selection is thus extended to the entire search space rather than to that implied by the immediate population. However, the *relative* performance of the GA in these two landscapes is extremely interesting. When the standard GA was applied originally, it did much better on *Onemax* than on its Gray-encoded version. Now, the 'headless chicken' GA can hardly distinguish between them— the computational growth rate is almost identical at about $(1.38)^l$ in the case of *Onemax*, and $(1.37)^l$ for the Gray-encoded version.

It is clear, therefore, that the GA benefits more from the traditional parent selection scheme when applied to the binary-encoded version than when applied to the Gray-encoded version. This can be explained if we consider the heuristic underlying the traditional parent selection scheme.

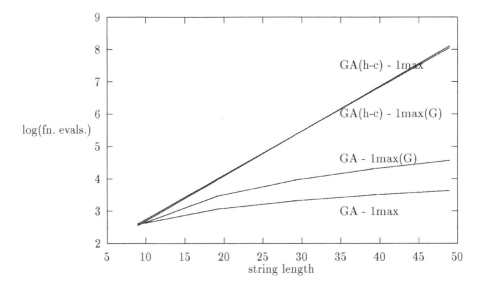

Fig. 3. Performance of the 'headless chicken' GA for *Onemax* and Gray-encoded *Onemax*. For comparison, the results of the original GA are also included. As before, the vertical axis measures the number of function evaluations needed (averaged over 30 independent trials), and is on a logarithmic scale.

In the 'headless chicken' experiments the distance between the selected parents is on average about $l/2$. Moreover it remains constant at this value throughout the search. In the traditional parent selection scheme, by contrast, the distance between parents is a function of the population composition, and it decreases as the population converges.

As remarked above, the distance between parents determines the size of the component to which the search is restricted. But it does more than this: it also changes the nature of the landscape. For example, if the parents are complementary, i.e. if they are at maximal Hamming distance, the 1-point crossover landscape is identical to the CX landscape. On the other hand, the minimal distance at which recombination shows any effect is given when the parents are at Hamming distance 2 from each other. In this case, the crossover landscape is a fragment of the BF landscape—in the sense that if the offspring is not a 'clone' of one of the parents, it must be Hamming distance 1 from both parents.

Consequently, adjusting the distance between parents makes the crossover landscapes 'look' more like the CX landscape, or more like the BF landscape. This gives the GA an ability to solve problems that are difficult in the CX landscape but easy in the BF landscape (such as the *Onemax* function). However, this process is asymmetrical. As the population converges the GA can only generate crossover landscapes that are similar to the BF landscape. Both theory and experiment show that Gray-encoded *Onemax* is hard for a hill-climber in the BF landscape. Thus pairwise mating (although still beneficial) has less of an impact on the Gray-encoded version. However, for the 'headless chicken' GA all the crossover landscapes generated lie somewhere between the CX landscape and the BF landscape, and thus neither function is favoured.

5   Practical Applications

If we were to give a concise summary of the above analysis, it would be to stress that there is more to landscapes than their Walsh decompositions. While it is undeniably useful that we can construct techniques which help us neatly to summarize certain facts about a landscape, we must recognize that there are other features—possibly very important ones—that are not captured by these methods. In the simple example of the cubic function we have seen that the search strategy adopted can make a big difference to the likelihood of a NS hill-climber finding the global optimum. In the much more complex case of the *Onemax* function, again we have seen that the particular form of the search strategy adopted by a GA is crucial to its success—but this is not something that could be predicted merely from an analysis of the epistasis variance.

It may therefore be more important in practice to take note of the results of empirical studies of landscapes. One of the most interesting characteristics has been seen in many different studies: it is a feature of the well-known 'NK-landscapes' formulated by Kauffman [8], and it also appears in many examples of COPs, such as the travelling salesman problem or TSP [9,10], graph partitioning [11], and flowshop scheduling [12].

In the first place, such studies have repeatedly found that, on average, local optima are very much closer to the global optimum than are randomly chosen points, and closer to each other than random points would be. That is, the distribution of local optima is not *isotropic*, rather they are clustered in a 'central massif' (or—if we are minimizing—a 'big valley'). This can be demonstrated graphically by plotting a scatter graph of fitness against distance to the global optimum. Secondly, if the basins of attraction of each local optimum are explored, size is quite highly correlated with quality: the better the local optimum, the larger is its basin of attraction.

Of course, there is no guarantee that this property holds in any particular case, but it provides an explanation for the success of 'perturbation' methods [13,14,15] which currently appear to be the best available for the TSP. It is also tacitly assumed by such methods as simulated annealing and tabu search, which would lose a great deal of their potency if local optima were isotropically distributed. Recent work on path-tracing algorithms (reviewed in [16,17]) also rests on the assumption that the big valley conjecture is true.

It also suggests a starting point for the development of new ways of implementing evolutionary algorithms. The details have already been published elsewhere [18,19], so a simple sketch will suffice here. If we consider the case of crossover of vectors in $\mathbb{Z}_2^l$, it is easily seen that any 'child' produced from two 'parents' will lie on a path that leads from one parent to another. Figure 4 demonstrates this fact.

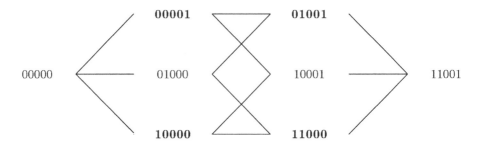

Fig. 4. The diagram shows the set of paths that could be traced between the parents 00000 and 11001. Only those intermediate vectors indicated by bold type can be generated by one-point crossover, but all can be generated by uniform crossover.

In an earlier paper [20], we described such points as 'intermediate vectors': intermediate in the sense of being an intermediate point on the Hamming landscape. In the case of non-binary strings, the distance measure may be more complicated, but the principle is still relevant. Thus crossover is re-interpreted as finding a point lying 'between' 2 parents in some landscape in which we hope the big valley conjecture is true. We implemented this 'path-tracing crossover' for both the makespan and the flowsum versions of the flowshop sequencing problem; Figure 5 shows in a 2-dimensional diagram the idea behind it, while full details can be found in [16,18].

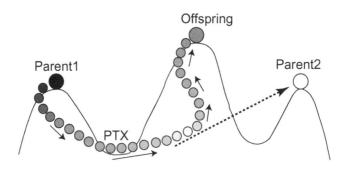

Fig. 5. Path tracing crossover combined with local search: a path is traced from one parent in the direction of the other. In the 'middle' of the path, solutions may be found that are not in the basins of attractions of the parents. A local search can then exploit this new starting point by climbing to the top of a hill (or the bottom of a valley, if it is a minimization problem)—a new local optimum. The acronym PTX signifies 'path-tracing crossover'.

In this way, the concept of recombination can be fully integrated with traditional NS methods, and the results obtained for flowshop instances (see [18,19] for details) were gratifyingly good. For the makespan problem, embedded path tracing helped the GA to achieve results of outstandingly high quality: several new best solutions were discovered for well-known benchmarks. For the flowsum version, optimal solutions are not known, but the path-tracing GA consistently produced better solutions than other proposed techniques.

6 Conclusion

This paper has reviewed and discussed in some detail the basic mathematical theory and methods associated with the concept of a search *landscape*. While these methods can be very useful in enhancing our understanding of evolutionary algorithms, it has been emphasized that they cannot provide a complete explanation for the performance of a specific algorithm on their own—even (as discussed here) in the case of very simple functions. Secondly, and more briefly, some empirically determined properties of many search landscapes have been described, and one approach whereby such properties can be exploited has been outlined.

Finally, we should remark that several interesting future research questions are suggested. For instance, can we provide a formal definition of what it means for a 'big valley' structure to exist, and can we relate it to mathematical constructs associated with neighbourhood structures? Does the big valley exist almost everywhere? If not, can we define classes of problems for which it does not occur? In the area of implementation, how can we further refine the path tracing methodology and its integration into evolutionary algorithms?

As our understanding of the nature of search landscapes and how to exploit them develops, this promises to become an important area of research into the theory and application of evolutionary algorithms.

References

1. C.Höhn and C.R.Reeves (1996) The crossover landscape for the *onemax* problem. In J.Alander (Ed.) (1996) *Proceedings of the 2nd Nordic Workshop on Genetic Algorithms and their Applications*, University of Vaasa Press, Vaasa, Finland, 27-43.
2. T.C.Jones (1995) *Evolutionary Algorithms, Fitness Landscapes and Search*, Doctoral dissertation, University of New Mexico, Albuquerque, NM.
3. L.K.Grover (1992) Local search and the local structure of NP-complete problems. *Operations Research Letters*, **12**, 235-243.
4. P.F.Stadler and G.P.Wagner (1998) Algebraic theory of recombination spaces. *Evolutionary Computation*, **5**, 241-275.
5. C.R.Reeves (1999) Predictive measures for problem difficulty. In *Proceedings of 1999 Congress on Evolutionary Computation*, IEEE Press, 736-743.
6. J.C.Culberson (1995) Mutation-crossover isomorphisms and the construction of discriminating functions. *Evolutionary Computation*, **2**, 279-311.

7. T.C.Jones (1995) Crossover, macromutation and population-based search. In L.J.Eshelman (Ed.) (1995) *Proceedings of the 6th International Conference on Genetic Algorithms*, Morgan Kaufmann, San Francisco, CA, 73-80.

8. S.Kauffman (1993) *The Origins of Order: Self-Organization and Selection in Evolution.* Oxford University Press.

9. S.Lin (1965) Computer solutions of the traveling salesman problem. *Bell Systems Tech. J.* ,**44**,2245-2269.

10. K.D.Boese, A.B.Kahng and S.Muddu (1994) A new adaptive multi-start technique for combinatorial global optimizations. *Operations Research Letters*, **16**, 101-113.

11. P.Merz and B.Freisleben (1998) Memetic algorithms and the fitness landscape of the graph bi-partitioning problem. In A.E.Eiben, T.Bäck, M.Schoenauer, H-P.Schwefel (Eds.) (1998) *Parallel Problem-Solving from Nature—PPSN V*, Springer-Verlag, Berlin, 765-774.

12. C.R.Reeves (1999) Landscapes, operators and heuristic search. *Annals of Operational Research*, **86**, 473-490.

13. D.S.Johnson (1990) Local optimization and the traveling salesman problem. In G.Goos and J.Hartmanis (Eds.) (1990) *Automata, Languages and Programming*, Lecture Notes in Computer Science **443**, Springler-Verlag, Berlin, 446-461.

14. O.Martin, S.W.Otto and E.W.Felten (1992) Large step Markov chains for the TSP incorporating local search heuristics. *Operations Research Letters*, **11**, 219-224.

15. G.Zweig (1995) An effective tour construction and improvement proedure for the traveling salesman problem. *Operations Research*, **43**, 1049-1057.

16. C.R.Reeves and T.Yamada (1999) *Embedded Path Tracing and Neighbourhood Search Techniques in Genetic Algorithms.* In K.Miettinen, M.M.Mäkelä, P.Neittaanmäki and J.Périaux (Eds.) (1999) *Evolutionary Algorithms in Engineering and Computer Science*, John Wiley & Sons, Chichester, 95-111.

17. C.R.Reeves and T.Yamada (1999) *Goal-Oriented Path Tracing Methods.* To appear in D.A.Corne, M.Dorigo and F.Glover (Eds.) (1999) *New Methods in Optimization*, McGraw-Hill, London.

18. C.R.Reeves and T.Yamada (1998) Genetic algorithms, path relinking and the flowshop sequencing problem. *Evolutionary Computation*, **6**, 45-60.

19. T.Yamada and C.R.Reeves (1998) Solving the C_{sum} permutation flowshop scheduling problem by genetic local search. In *Proc. of 1998 International Conference on Evolutionary Computation*, 230–234, IEEE Press.

20. C.R.Reeves (1994) Genetic algorithms and neighbourhood search. In T.C.Fogarty (Ed.) (1994) *Evolutionary Computing: AISB Workshop, Leeds, UK, April 1994; Selected Papers.* Springer-Verlag, Berlin, 115-130.

On the Effectivity of Evolutionary Algorithms for the Multidimensional Knapsack Problem

Jens Gottlieb

Department of Computer Science, Technical University of Clausthal
Julius-Albert-Str. 4, 38678 Clausthal-Zellerfeld, Germany
`gottlieb@informatik.tu-clausthal.de`

Abstract. When designing evolutionary algorithms (EAs) for the multidimensional knapsack problem, it is important to consider that the optima lie on the boundary B of the feasible region of the search space. Previously published EAs are reviewed, focusing on how they take this into account. We present new initialization routines and compare several repair and optimization methods, which help to concentrate the search on B. Our experiments identify the best EAs directly exploring B.

1 Introduction

The *multidimensional knapsack problem (MKP)* is stated as

$$\text{maximize} \quad \sum_{j \in J} p_j x_j \tag{1}$$

$$\text{subject to} \quad \sum_{j \in J} r_{ij} x_j \le c_i, \quad i \in I \tag{2}$$

$$x_j \in \{0, 1\}, \qquad j \in J \tag{3}$$

with $I = \{1, \ldots, m\}$ and $J = \{1, \ldots, n\}$ denoting the sets of resources and items, respectively. Each item j has a profit $p_j > 0$ and a resource demand $r_{ij} \ge 0$ of resource i, which is limited by its capacity $c_i > 0$. The goal is to determine a set of items with maximum profit, which does not exceed the resource capacities.

The MKP models many economic problems and is NP-complete [GJ79], hence many heuristics have been proposed in literature [CB98]. When applying evolutionary algorithms (EAs) to constrained problems like MKP, the most important issue is the choice of an appropriate constraint handling technique [Mic95]. We demonstrate that a careful analysis of the MKP structure can explain the effectivity of EAs. This paper extends our previous work [Go99], presenting new results concerning the effects of initialization methods, ordering heuristics, their degree of nondeterminism, and the influence of duplicate elimination.

We proceed as follows. The relevant structural properties of MKP are provided in Sect. 2, followed by a literature review of EAs for MKP in Sect. 3. Section 4 presents procedures for initialization, repair and optimization together with a detailed comparison of several combined approaches. Finally, the conclusions and ideas for further research are given in Sect. 5.

C. Fonlupt et al. (Eds.): AE'99, LNCS 1829, pp. 23–37, 2000.

2   The Boundary of the Feasible Region

The goal is to determine $x \in S = \{0,1\}^n$ which maximizes profit and satisfies the resource constraints. The *search space* S is partitioned into the *feasible region*

$$F = \{x \in S \mid \sum_{j \in J} r_{ij}x_j \le c_i \text{ for all } i \in I\},$$

and the *unfeasible region* $U = S \setminus F$. We assume the most natural neighbourhood for $x \in S$, i.e. $N(x) = \{y \in S \mid d(x,y) = 1\}$ for Hamming distance $d : S \times S \to \mathbb{N}$. The relation $\prec$ over $S \times S$ is defined as follows: $x \prec y$ iff $y \in N(x)$ and $x_j < y_j$ for some $j \in J$. Assuming $x \prec y$, the following statements are easy to verify: (i) $y \in F$ implies $x \in F$, (ii) $x \in U$ implies $y \in U$, and (iii) y yields a higher profit than x. As a consequence of (i) and (ii), both F and U are connected with respect to the transitive closure of the neighbourhood. Furthermore suppose $y \in F$, then the set $X = \{x \in S \mid x \prec y\}$ is a subset of F but does not contain any global optimum, due to (i) and (iii). Hence large parts of the feasible region cannot contain a global optimum. Another interpretation exhibits the fact that for a global optimum $x^* \in F$ the set $Y = \{y \in S \mid x^* \prec y\}$ is unfeasible ($Y \subseteq U$). This helps us to identify the most promising part of the feasible region, the *boundary* B of the feasible region, which represents that part of F neighbouring to unfeasible solutions, i.e.

$$B = \{x \in F \mid x \prec y \text{ implies } y \in U\}.$$

The introduced concepts and properties are demonstrated in Fig. 1. Above considerations imply that all global optima are contained in B. Any reasonable search algorithm must ensure that a global optimum is among the potential solutions, hence the sets S, F and B are candidates to be searched, while U is not a candidate. As $B \subseteq F \subseteq S$ holds, random sampling of B has a higher probability to find the global optimum than random sampling of F, which itself dominates random sampling of S. It is interesting to check whether similar relations hold for EAs. Therefore we call an EA *biased towards* B, if it focuses the search on B, and examine EAs how their performance is affected by this bias towards B.

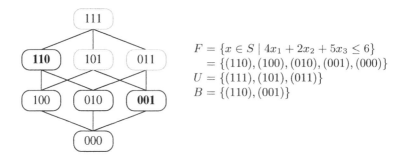

$$F = \{x \in S \mid 4x_1 + 2x_2 + 5x_3 \le 6\}$$
$$= \{(110), (100), (010), (001), (000)\}$$
$$U = \{(111), (101), (011)\}$$
$$B = \{(110), (001)\}$$

Fig. 1. Example for $S = \{0,1\}^3$, the relation $\prec$, and F, U, B

3   Evolutionary Algorithms for MKP

We classify EAs for MKP by the search space which is explored by the population, hence S, F and B represent the alternatives.[1] EAs are also able to solve MKP via searching A, an arbitrary set, if there is a decoder available which maps A into an original MKP search space S, F or B. We call such an approach *indirect*, while EAs operating in original search spaces are called *direct*. As bit strings directly match S, classical genetic algorithms (GAs) [Hol75] with standard bit mutation, one-point crossover, fitness proportional parent selection and generational replacement are often used. To present an overview of relevant EAs, we include two special cases of the MKP, namely the *(unidimensional) knapsack problem (KP, $m = 1$)* and the *subset sum problem (SSP, $m = 1$ and $p_j = r_{1j}$)*. Moreover we also briefly mention some papers, which do not focus on constraint handling techniques, but instead mainly examine other components or extensions of EAs (and use MKP as test case).

3.1   Direct Search in S

Direct search in S means that both feasible and unfeasible solutions contribute to the search process. The evaluation of unfeasible solutions is usually based on penalties, which degrade the fitness of such solutions. In the case of SSP, Khuri et al. used a penalty linear in the distance from feasibility, containing an offset term to ensure that the best unfeasible candidate can never be better than the worst feasible one [KBH94a]. Hence this approach concentrates on the feasible region F, deteriorating the influence of unfeasible individuals.

A penalty function measuring the ratio of the constraint violation to the sum of all resource demands was proposed by Khuri and Batarekh for the KP [KB90]. Thiel and Voss adapted this penalty function to the MKP and reported poor results for this approach, which sometimes even had problems in finding any feasible solution [TV94]. Michalewicz and Arabas employed logarithmic, linear and quadratic penalty functions for the KP, depending on the amount of constraint violation and allowing unfeasible candidates to beat bad feasible candidates [MA94]. They observed all three penalty approaches failing to find any feasible solution for instances with a low ratio $|F|/|S|$ [MA94]. This was also recognized by Olsen, who experimented with a set of strong, moderate and weak penalty functions for the KP [Ols94]. She reported best results for strong penalties depending on profit and distance from feasibility, while weak penalties prevented the GA from finding any feasible solution for low $|F|/|S|$. Olsen also tried penalties based on evolution time only, but observed their inferiority. Gordon et al. examined different parallel GAs for KP, proposing a penalty equal to the constraint violation [GBW93,GW93].

Some penalty-based approaches for MKP were investigated by Khuri et al. [KBH94b] and Hoff et al. [HLM96], who used a steady-state GA and uniform

---

[1] Thus, the search space of an EA, which repairs solutions $x \in U$, is F since each population is feasible.

crossover. Hoff et al. reported a penalty defined by the square root of the sum of the constraint violations superior to the penalty function of [KBH94b], which depends on the number of violated constraints, and felt Khuri et al.'s penalty could be too strong. We suppose another reason for its inferiority is that the amount of each constraint violation is ignored (according to the early findings of Richardson et al. concerning the design of penalties [RPLH89]). However, the selfish gene algorithm [CSRS98], the cellular GA with self-adapting acceptance threshold [RS95,RS96], and the ptGA, which is based on noncoding segments and promoter/terminator sequences [May98], used the penalty of [KBH94b] for MKP. Lin et al. proposed combining simulated annealing with a GA for MKP and penalizing unfeasible individuals [LKH93].

Generally, penalty-based EAs are quite sensitive to the problem structure at hand. Some authors proposed alternative initialization routines [TV94,KBH94b, HLM96], because their EAs failed to find any feasible solution in some cases, e.g. caused by a low ratio $|F|/|S|$. Hence there is a need for a bias towards F, which can be achieved by strong penalties depending on the distance to feasibility, favouring unfeasible solutions near the boundary B. Thus the bias towards B strongly affects the success of penalty-based EAs.

3.2   Direct Search in F

Among the alternatives to restrict the search to F are specialized operators ensuring feasibility and repair algorithms, which are very common in EAs for MKP and alter a previously unfeasible solution (produced by crossover and mutation) into a feasible one.

Michalewicz and Arabas used a general repair method for KP, which removes items from the current unfeasible solution until the knapsack constraint is satisfied [MA94]. Their algorithm is parameterized by the order in which the items are processed. A comparison of random and greedy ordering exhibited the clear superiority of the greedy approach, which tries to repair optimally. That result coincides with experiments, in which we observed the high performance of the simple greedy algorithm (see [MT90]) for KP. Michalewicz and Arabas also examined whether it is useful to replace only some unfeasible solutions by their repaired version, and use the repaired version only for evaluation purposes. However, replacing seems not to have an impact for any replacement percentage considered [MA94]. The random repair method was applied by Hoff et al. to MKP, yielding a better performance as for the considered penalty functions [HLM96] and a comparable performance as a tabu search method [Løk95]. Yang compared line-breeding schemes for MKP [Yan98], relying on operators which ensure feasibility and are conceptually similar to random repair methods. Zitzler and Thiele adapted the greedy repair method to a multiobjective version of KP and compared several multiobjective EAs in this context [ZT98]. In their studies concerning parallel GAs for KP [GBW93,GW93], Gordon et al. also considered a repair method working similar as greedy repairing.

Thiel and Voss compared several approaches for MKP reporting best results for an EA with heuristic initialization, repair and optimization operators [TV94].

While all unfeasible individuals are repaired, the feasible solutions are improved by the optimization method, which ensures reaching the boundary B and is applied with a low probability. Thiel and Voss reported worse results for higher probabilities, which might be caused by a loss of diversity due to the optimization heuristic. They also proposed a hybridization with tabu search, which improved the obtained quality but caused significant higher CPU times. The best results for MKP were reported for the EAs of Chu and Beasley [Chu97,CB98] and Raidl [Rai98], who also used heuristic repair, optimization and initialization routines. Both EAs are strongly biased towards B due to the optimization method, however the initialization may also produce individuals $x \in F \setminus B$. A detailed description of their heuristic methods is presented in Sect. 4, together with ideas for improvement.

In general, the restriction to F allows to obtain better results than for most penalty-based EAs. Moreover this approach is more flexible, i.e. its success does not depend so strongly on the problem structure as penalty-based algorithms do. While repair methods produce solutions near the boundary, the strongest bias towards B is achieved by combinations with optimization routines.

3.3   Indirect Search

Michalewicz and Arabas proposed a decoder for KP, based on the ordinal representation, and investigated two variants based on random and greedy decoding [MA94]. Both decoders are mappings onto B, nevertheless they were dominated by the considered direct methods. The reason should be the missing locality, i.e. a small difference in genotype might result in an extremely different phenotype.

Hinterding examined an order based EA for KP, using a first fit algorithm to decode a permutation of items into a feasible solution [Hin94]. This approach can easily be extended to solve MKP, which was done by Raidl for comparison purposes [Rai98]. Thiel and Voss applied this approach to MKP and compared different crossover operators, but the results were inferior to some direct methods [TV94]. Hinterding also introduced a variable-length representation, which resembles a selection of items fitting into the knapsack [Hin94]. He calls this representation a direct encoding, however there exist many representations of the same KP solution (the order of the items can be arbitrarily permuted), which is typical for decoders. Furthermore the employed injection crossover and in particular the mutation operator make explicit use of the item ordering when using the first fit algorithm, hence we feel this approach could rather be classified as an indirect search method. The variable-length representation is inferior on the larger problem examined, which might be caused by the fact that the order based decoder produces only solutions $x \in B$, while for the variable-length representation mutation and crossover may produce candidates $x \in F \setminus B$ (however, the initialization ensures to generate candidates $x \in B$).

Recently, Raidl proposed a decoder approach based on real-valued weights for the items, which are used to modify the objective function (1) yielding a similar but slightly different MKP instance [Rai99]. This problem is then solved by a simple heuristic resulting in a solution candidate for the original MKP, which is feasible since the resource constraints (2) remain unchanged. Raidl compared different biasing techniques of the original MKP, and two decoding heuristics which ensure to generate solutions $x \in B$. The weight-coding approach yields better results than other indirect approaches, due to the strong locality and heuristic bias [Rai99,GR99]. However, the approach is slightly inferior to the best direct approaches [Chu97,CB98,Rai98] with respect to solution quality.

Although indirect methods were proposed which search in B, most of them suffer from a lack of locality. Hence, the bias towards B is important but cannot guarantee good performance for indirect approaches.

4   Direct Search in B

While some papers present indirect methods searching in B, only few attempts have been made to explicitly search the boundary by a direct method. Hence our aim is to identify an EA setup that directly explores B, therefore we choose the currently best EAs [Chu97,CB98,Rai98] as starting point for our investigations. After introducing the general setup of our experiments, we compare different initialization routines and examine repair and optimization methods.

4.1   General Setup

All computational results are based on the MKP test suite introduced by Chu [Chu97] and available from the OR-Library[2]. The instances are grouped into nine different problem sizes ($m \in \{5, 10, 30\}$ and $n \in \{100, 250, 500\}$), each group containing 30 instances with three different ratios $|F|/|S|$, yielding a total of 270 instances. We choose the same general EA configuration as Chu and Beasley [Chu97,CB98] and Raidl [Rai98]: population size 100, parent selection via tournaments of size 2, steady-state replacement scheme (replacing the worst individual), duplicate elimination and a limit of 1 000 000 non-duplicate evaluations. Furthermore, we use the same operators as Raidl, namely uniform crossover with probability $p_c = 0.9$ and standard bit mutation with $p_m = 1/n$ (Chu and Beasley use the same operators, but $p_c = 1.0$).

We measure the obtained quality by the *percentage gap* $100(1 - max^{EA}/opt^{LP})$, which is an upper bound of the distance from the optimum. The best objective value found by an EA run is denoted by max^{EA}, while opt^{LP} is the optimal value of the linear programming (LP) relaxation obtained from MKP by relaxing the integrality constraints (3).

4.2   Initialization Routines

We describe two routines from literature to generate a solution candidate, and suggest simple improvements for these procedures. All considered routines are based on a random permutation $\pi : J \rightarrow J$ of the items and start with the

---

[2] http://mscmga.ms.ic.ac.uk/info.html

solution candidate $x = (0, \ldots, 0)$. Most important is the procedure $\mathtt{try}(x, j)$, which increases x_j from 0 to 1 if the resulting solution is feasible.

The method proposed by Chu and Beasley [Chu97,CB98] – in the following abbreviated by C – traverses the items in the order determined by π, calling $\mathtt{try}(x, \pi(j))$ for each $\pi(j)$ and stopping as soon as the current variable $x_{\pi(j)}$ was not increased by $\mathtt{try}$. Obviously, the resulting solution could be improved if the remaining part of the items would also be checked for inclusion. The routine C* based on that idea checks *all* items via $\mathtt{try}$ and does *not terminate* if the first item is found which cannot be included. While C may produce solutions in $F \setminus B$, C* ensures generating only solutions $x \in B$. Note that C* corresponds to the initialization routine for the order based encoding from Hinterding [Hin94].

Raidl proposes a method which makes use of the LP optimal solution x^{LP}, based on the intuition that higher x_j^{LP} values indicate a better usefulness of setting $x_j = 1$ [Rai98]. The procedure – denoted by R – proceeds by traversing all variables in the order determined by π and calling $\mathtt{try}(x, \pi(j))$ with probability $x_{\pi(j)}^{LP}$. Hence, the promising items have a higher probability for inclusion, but the produced solution x might fail to reach the boundary B. Obviously, R can also be improved in a fashion similar to C*, resulting in method R*. The first phase of R* uses R to produce an initial solution, and the second phase ensures reaching B by traversing all items in the order π, which were not selected for a $\mathtt{try}$ call in the first phase, and calling $\mathtt{try}(x, \pi(j))$ for each. While items j with $x_j^{LP} = 0$ are ignored by R, R* might select such items for inclusion, too.

4.3   Comparing the Initialization Routines

The four initialization routines described in Sect. 4.2 have been tested in 100 runs for each instance, yielding 3000 runs for each problem size, and are compared in Table 1 concerning different measures. The obtained quality is represented by the percentage gap (see Sect. 4.1). While the *average Hamming distance* characterizes the population diversity, the *average Hamming weight* can be used as a measure for the distance of the population from B: A higher value indicates a smaller average distance to B. It should be remarked that only similar methods should be compared by this measure in this sense, since the boundary usually contains solutions with different Hamming weights (see Fig. 1) and different methods might produce solutions at different Hamming weight levels.

We have also calculated the *boundary ratio* (not shown in Table 1), which measures the fraction of the population that is contained in the boundary B. While C and R yield a boundary ratio 0.05 and 0.47 respectively, it is clear that C* and R* achieve value 1. Both methods C* and R* yield a better quality than their counterparts C and R, which can be explained by the much higher boundary ratio of C* and R*. This coincides with the lower average Hamming weight of C and R, implying a higher distance from B, i.e. the resources are not used completely. We also observe a higher quality improvement of C* with respect to C as for R* and R. One reason is that C fails to reach the boundary for most individuals, while the underlying heuristic enables R to produce much more individuals on B. Furthermore, R already exploits promising items in the

Table 1. Comparison of the initialization routines C, C*, R and R*

		percentage gap				avg. Hamming distance				avg. Hamming weight			
m	n	C	C*	R	R*	C	C*	R	R*	C	C*	R	R*
5	100	14.98	12.37	1.22	0.92	37.2	37.5	5.0	5.8	46.4	48.5	51.6	52.2
5	250	14.87	13.40	0.47	0.32	93.8	94.1	5.2	6.3	119.6	122.4	131.0	131.7
5	500	15.04	14.22	0.23	0.15	187.6	187.6	5.1	6.5	242.7	245.8	264.1	265.0
10	100	15.55	13.30	1.73	1.52	37.1	37.4	6.8	7.5	45.5	47.2	50.0	50.4
10	250	15.42	13.93	0.70	0.58	93.7	94.0	7.1	8.1	118.0	120.6	127.0	127.5
10	500	15.12	14.24	0.36	0.28	187.4	187.6	7.4	8.6	240.5	243.6	255.7	256.4
30	100	16.58	14.47	3.12	2.63	37.0	37.3	10.7	11.3	44.2	45.7	47.2	47.7
30	250	16.04	14.74	1.44	1.21	93.6	93.9	13.4	14.3	116.0	118.3	122.5	123.1
30	500	15.63	14.80	0.76	0.64	187.2	187.5	14.8	15.9	237.5	240.3	249.3	245.0

first phase so that only few good items remain for the improvement steps of R*. In the case of C, there are more promising items left which can be exploited by the second phase of C*.

Both R and R* exhibit a clearly better quality than C and C*, since the LP heuristic prefers items with high profit and low resource demands. This results in selecting more profitable items and including more items (see the average Hamming weight), due to the more efficient use of resources. These profitable items have an interesting effect on the average Hamming weight: R* and even R produce solutions with higher Hamming weight than C*, although C* guarantees to reach B and R has a much lower boundary ratio.

Concerning the average Hamming distance, we observe a difference of an order of a magnitude between C and C* on the one hand, and R and R* on the other hand. While the populations produced by C and C* are very wide-spread, R and R* yield a concentration on promising but relatively similar solution candidates. Items j with $x_j^{LP} = 1$ and $x_j^{LP} = 0$ have a very high and low probability of being selected, respectively, and hence limit the main variations in the population to the items j with fractional x_j^{LP}. The overall effect of the initial diversity depends on the used operators, since a high initial diversity allows the use of more deterministic operators, while a low initial diversity needs nondeterministic operators to prevent premature convergence. Interestingly, the improved routines exhibit an increase of population diversity. While for C* this increase is negligible, R* yields a relatively high increase of the average Hamming distance, since the second phase of R* uses a random permutation and hence introduces additional variations. Generally, the enhanced procedures yield a better quality *and* population diversity, therefore C* and R* should be favoured.

4.4 Repair and Optimization Methods

As we chose uniform crossover and standard bit mutation as operators, it is probable to obtain an offspring $x \in F \setminus B$ or $x \in U$. In order to focus the search on B additional operators must be used. As mentioned in Sect. 3.2, solutions $x \in U$ can be made feasible by repair methods. Most repair methods iteratively employ a procedure $\mathtt{reset}(x, j)$, which sets $x_j = 0$. The methods from [Chu97, CB98,Rai98] are based on a permutation $\pi : J \to J$ and proceed by traversing all items j with $x_j = 1$ in the order π, calling $\mathtt{reset}(x, \pi(j))$ and terminating as soon as the current solution x has become feasible. Two approaches for determining the order π have been proposed. Chu and Beasley calculate a *pseudo-utility ratio* for each item (based on shadow prices of the constraints of the LP relaxation) and determine π by sorting the items accordingly [Chu97,CB98]. Raidl proposes to use the *LP optimum* x^{LP} as sorting criterion to obtain π and suggests to randomly shuffle all items with the same x_j^{LP} value [Rai98], which introduces more nondeterminism into π since there exist many items j with $x_j^{LP} \in \{0, 1\}$.[3] When repairing an individual, the least promising items are processed first to retain the promising items in the solution.

Repairing unfeasible individuals ensures reaching F but not B. Therefore, feasible individuals must be optimized to achieve solutions on the boundary. While we could restrict the EA to optimize repaired individuals only, Chu, Beasley and Raidl proposed to optimize all individuals generated by crossover and mutation (and repaired, if necessary). The optimization is also based on an item ordering π and is conceptually similar to the initialization routines. The current solution x is improved by traversing all items j with $x_j = 0$ in the order π and calling $\mathtt{try}(x, \pi(j))$ (see Sect. 4.2) for each of them. Contrary to the repair algorithm, the optimization processes the items in the reverse order, meaning that promising candidates are checked for possible inclusion first and (probably) worse items last. Note that this method always produces a solution $x \in B$, independent of the used permutation π.

It must be remarked that Chu and Beasley never change the original order π, yielding a very effective deterministic repair and optimization procedure. On the other hand, Raidl shuffles the order before repairing *and* before optimizing, which yields an effective nondeterministic repair and optimization method. To investigate the effects of nondeterminism more detailed, we generalize the proposed order heuristics (i.e. pseudo-utility ratio, LP optimum) in the following way. Each time when an operator (repair or optimization) needs a permutation π, the heuristic order is used with probability p_h and a randomly generated permutation is employed otherwise. While $p_h = 1$ reflects the highest influence of the heuristic, $p_h = 0$ resembles completely random repair and optimization.

4.5   Comparing the Repair and Optimization Methods

Tables 2 and 3 compare different approaches, considering the quality for each problem size – also averaged over all problems – and the average number of evaluations needed to find the best solution (AEBS), relative to the number of all generated non-duplicate solutions. Three runs were considered for each

---

[3] More precisely, there exist at least $n - m$ such variables since only the m basic variables might be fractional.

Table 2. Obtained quality for pseudo-utility ratio based approaches

		C*, OAI, different p_h values					R*, OAI, different p_h values					C, $p_h = 1$	
m	n	0	0.25	0.5	0.75	1	0	0.25	0.5	0.75	1	OAI	ORI
5	100	0.587	0.587	0.587	0.587	0.586	0.587	0.586	0.586	0.587	0.587	0.587	0.587
5	250	0.179	0.153	0.147	0.146	0.142	0.166	0.152	0.149	0.145	0.143	0.140	0.143
5	500	0.120	0.059	0.055	0.055	0.053	0.069	0.058	0.055	0.054	0.053	0.053	0.055
10	100	0.974	0.957	0.962	0.964	0.967	0.967	0.962	0.958	0.965	0.965	0.961	0.985
10	250	0.397	0.319	0.318	0.306	0.308	0.340	0.324	0.316	0.306	0.301	0.305	0.316
10	500	0.294	0.153	0.145	0.143	0.140	0.162	0.149	0.148	0.144	0.140	0.141	0.144
30	100	1.718	1.713	1.702	1.701	1.701	1.719	1.703	1.709	1.701	1.704	1.709	1.712
30	250	0.790	0.694	0.680	0.684	0.673	0.718	0.698	0.687	0.682	0.675	0.676	0.680
30	500	0.552	0.385	0.362	0.358	0.359	0.372	0.361	0.358	0.354	0.354	0.351	0.358
total		0.624	0.558	0.551	0.549	0.548	0.567	0.555	0.552	0.549	0.547	0.547	0.553
AEBS		0.527	0.324	0.289	0.275	0.253	0.333	0.285	0.273	0.270	0.261	0.257	0.292

instance, yielding 90 runs for each problem size and 810 runs for each EA. The approaches are characterized by their order heuristic (pseudo-utility ratio, LP optimum), initialization routine (C, C*, R, R*), optimization strategy (optimize all individuals, optimize repaired individuals) and degree of nondeterminism (p_h). Note that the original EA from Chu and Beasley uses the pseudo-utility ratio and C, while Raidl's EA employs the LP optimum and R; both EAs are based on $p_h = 1$ and optimizing all individuals.

The strategy to optimize all individuals (OAI) is superior to optimizing only repaired individuals (ORI), caused by the higher influence of the optimization routine and hence the higher bias towards B. Since this is quite intuitive, we discard ORI from further consideration and concentrate on OAI.

Generally, random repair and optimization ($p_h = 0$) yields the worst results. Interestingly, the results for R* are better than for C* in the case $p_h = 0$. This can be explained by the heuristic used by R*, indicating that the quality of the initial population is influential but the initial diversity is not. Higher p_h values tend to degrade the influence of the initial quality and also yield better results. We observe a clear dependency between p_h and achieved performance in terms of both quality and AEBS, which indicates the importance of the heuristic ordering.

Besides above results concerning different initialization methods we also observe that for $p_h = 1$ the routines C and C* perform quite similar, as R and R* do too. Generally, the influence of initialization routines is small if an appropriate optimization heuristic is applied.

For higher p_h values, LP optimum based EAs yield results inferior to pseudo-utility ratio based ordering. On the one hand, ordering according to the LP optimum does not distinguish between variables with same x_j^{LP} value, hence random ordering of such variables is neccessary to prevent a bias towards some of them. On the other hand, a fine grained distinction is achieved by the pseudo-utility

Table 3. Obtained quality for LP optimum based approaches

		C*, OAI, different p_h values					R*, OAI, different p_h values					R, $p_h = 1$	
m	n	0	0.25	0.5	0.75	1	0	0.25	0.5	0.75	1	OAI	ORI
5	100	0.588	0.588	0.587	0.586	0.588	0.587	0.587	0.586	0.586	0.586	0.586	0.591
5	250	0.184	0.157	0.153	0.154	0.150	0.168	0.156	0.155	0.151	0.150	0.152	0.155
5	500	0.116	0.063	0.061	0.061	0.059	0.067	0.062	0.061	0.060	0.058	0.059	0.063
10	100	0.971	0.973	0.970	0.968	0.976	0.963	0.969	0.962	0.969	0.968	0.975	0.986
10	250	0.386	0.331	0.326	0.320	0.316	0.336	0.329	0.323	0.321	0.318	0.320	0.327
10	500	0.288	0.157	0.151	0.152	0.148	0.162	0.156	0.153	0.148	0.146	0.149	0.154
30	100	1.712	1.705	1.712	1.705	1.710	1.722	1.708	1.701	1.710	1.702	1.711	1.727
30	250	0.786	0.686	0.694	0.683	0.676	0.722	0.701	0.691	0.684	0.680	0.678	0.687
30	500	0.566	0.370	0.362	0.362	0.357	0.374	0.370	0.366	0.358	0.358	0.359	0.364
total		0.622	0.559	0.557	0.555	0.553	0.567	0.560	0.555	0.554	0.552	0.554	0.562
AEBS		0.519	0.305	0.290	0.281	0.274	0.315	0.308	0.283	0.281	0.287	0.288	0.309

ratio, because only few variables are assigned the same priority. This explains the pseudo-utility ratio's superiority, coinciding with the similar performance for low p_h values, which degrade the influence of the used ordering heuristic.

4.6   The Effects of Duplicate Elimination

According to [Hin94,Chu97,CB98,Rai98], the importance of duplicate elimination should be emphasized, i.e. a newly generated solution is only accepted, if it is not already contained in the population. A loss of diversity and hence premature convergence can be prevented by this approach. However, the generation of too many duplicates might cause computational costs, which must not be neglected. We measure this effort by the *duplicate ratio*, which is calculated by $100(1 - \text{NDS/DS})$ with NDS and DS counting the number of generated non-duplicate and duplicate solutions, respectively.

The typical duplicate ratio of the EAs is depicted in Fig. 2, averaged over 30 instances and three runs per problem. EAs optimizing only repaired individuals exhibit a lower duplicate ratio than their counterparts which optimize all individuals, due to the less frequent use of the optimization. Among the approaches which optimize all individuals, the influence of the initialization is negligible for all p_h. Generally, a similar duplicate ratio for both C* and R* is achieved for LP optimum (LPO) based approaches, while in the case of pseudo-utility ratio (PUR) ordering R* produces more duplicates than C*. Higher p_h values cause a higher duplicate ratio, because a frequent use of the ordering heuristic restricts the search to promising items, which raises the probability to produce good but already existing solutions. However, even for $p_h = 1$ the duplicate ratio is acceptable. Generally, the initial population diversity is not as critical as might have been expected.

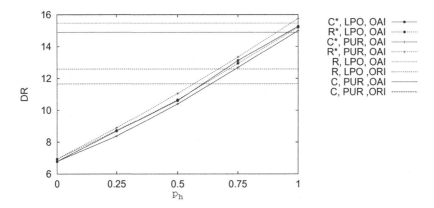

Fig. 2. Duplicate ratio for problem size $m = 10$ and $n = 500$

4.7   Summary

We conclude that the bias towards B does not guarantee good performance, but together with ordering heuristics the resulting direct EAs yield very good results, superior to all other EAs. Hence, the bias towards B can be interpreted as a basic requirement for a successful EA application. The best results are obtained for pseudo-utility ratio based ordering, optimizing all individuals, $p_h = 1$, and any initialization method C, C*, and R*.

It should be remarked that new best feasible solutions have been obtained during the EA runs for 114 instances, whose optima are not known, compared to the solutions obtained by Chu and Beasley [Chu97,CB98] and published at OR-Library. We omit a detailed comparison of CPU times for the different approaches, since they are conceptually equivalent (note that one iteration has the complexity $O(m \cdot n)$), coinciding with Raidl's observation concerning the computational efforts of his and Chu and Beasley's EA [Rai98]. For a comparison with other heuristic approaches regarding obtained quality and worst-case complexity, the reader is referred to [Chu97,CB98].

5   Conclusions and Further Research

We have presented an overview of EAs for the MKP, concentrating on the way they consider that all optima lie on the boundary B of the feasible region. The bias towards B has been used in an intuitive sense, hence there is a need for a formal approach, which enables a more accurate handling and also a more detailed comparison of existing constraint handling techniques. The measures we employed to compare the initialization routines could be extended from the static evaluation of the initial population to the characterization of the EA dynamics concerning the bias towards B. The – currently informal – concept of this bias helps to explain the success (or failure) of direct approaches, and is also essential for indirect algorithms, which strongly depend on the achieved locality.

Furthermore, the bias towards B is an important design issue, since all successful EAs for MKP are based on that principle.

There remain many interesting aspects to be studied more detailed, e.g. new ordering heuristics or alternative ways to control nondeterminism. Moreover, the gained insight should also be helpful to improve EAs for related problems.

References

[CB98]   P. C. Chu and J. E. Beasley. A Genetic Algorithm for the Multidimensional Knapsack Problem. *Journal of Heuristics.* Volume 4, No. 1, 63 – 86, 1998

[Chu97]   P. C. Chu. *A Genetic Algorithm Approach for Combinatorial Optimisation Problems.* PhD Thesis, The Management School, Imperial College of Science, Technology and Medicine, London, 1997

[CSRS98]   F. Corno, M. Sonza Reorda, and G. Squillero. The Selfish Gene Algorithm: a new Evolutionary Optimization Strategy. In *Proceedings of the 13th Annual ACM Symposium on Applied Computing.* 1998

[EBSS98]   A. E. Eiben, T. Bäck, M. Schoenauer, and H.-P. Schwefel (eds.). *Parallel Problem Solving from Nature - PPSN V.* Lecture Notes in Computer Science, Volume 1498, Springer, 1998

[GBW93]   V. Gordon, A. Böhm, and D. Whitley. *A Note on the Performance of Genetic Algorithms on Zero-One Knapsack Problems.* Technical Report CS-93-108, Department of Computer Science, Colorado State University, 1993

[GJ79]   M. R. Garey and D. S. Johnson. *Computers and Intractability: A Guide to the Theory of NP-Completeness.* W. H. Freeman, San Francisco, 1979

[Go99]   J. Gottlieb. Evolutionary Algorithms for Multidimensional Knapsack Problems: the Relevance of the Boundary of the Feasible Region. In *Proceedings of the Genetic and Evolutionary Computation Conference,* 787, 1999

[GR99]   J. Gottlieb and G. R. Raidl. Characterizing Locality in Decoder-Based EAs for the Multidimensional Knapsack Problem. In *Proceedings of Artificial Evolution,* 1999

[GW93]   V. Gordon and D. Whitley. Serial and Parallel Genetic Algorithms as Function Optimizers. In S. Forrest (ed.), *Proceedings of the Fifth International Conference on Genetic Algorithms,* 177 – 183, Morgan Kaufmann, 1993

[Hin94]   R. Hinterding. Mapping, Order-independent Genes and the Knapsack Problem. In *Proceedings of the 1st IEEE International Conference on Evolutionary Computation,* 13 – 17, 1994

[HLM96]   A. Hoff, A. Løkketangen, and I. Mittet. Genetic Algorithms for 0/1 Multidimensional Knapsack Problems. In *Proceedings Norsk Informatikk Konferanse,* 1996

[Hol75]   J. H. Holland. *Adaptation in Natural and Artificial Systems.* University of Michigan Press, Ann Arbor, 1975

[KB90]   S. Khuri and A. Batarekh. Heuristics for the Integer Knapsack Problem. In *Proceedings of the 10th International Computer Science Conference,* 161 – 172, Santiago, Chile, 1990

[KBH94a]   S. Khuri, T. Bäck, and J. Heitkötter. An Evolutionary Approach to Combinatorial Optimization Problems. In D. Cizmar (ed.), *Proceedings of the 22nd Annual ACM Computer Science Conference*, 66 – 73, ACM Press, New York, 1994

[KBH94b]   S. Khuri, T. Bäck, and J. Heitkötter. The Zero/One Multiple Knapsack Problem and Genetic Algorithms. In E. Deaton, D. Oppenheim, J. Urban, and H. Berghel (eds.), *Proceedings of the ACM Symposium on Applied Computation*, 188 – 193, ACM Press, 1994

[LKH93]   F.-T. Lin, C.-Y. Kao, and C.-C. Hsu. Applying the Genetic Approach to Simulated Annealing in Solving Some NP-Hard Problems. *IEEE Transactions on Systems, Man, and Cybernetics*. Volume 23, No. 6, 1752 – 1767, 1993

[Løk95]   A. Løkketangen. A Comparison of a Genetic Algorithm and a Tabu Search Method for 0/1 Multidimensional Knapsack Problems. In *Proceedings of the Nordic Operations Research Conference*, 1995

[MA94]   Z. Michalewicz and J. Arabas. Genetic Algorithms for the 0/1 Knapsack Problem. In Z. W. Raś and Z. Zemankova (eds.), *Proceedings of the 8th International Symposium on Methodologies for Intelligent Systems*, 134 – 143, Lecture Notes in Artificial Intelligence, Volume 869, Springer, 1994

[May98]   H. A. Mayer. ptGAs - Genetic Algorithms Evolving Noncoding Segments by Means of Promoter/Terminator Sequences. *Evolutionary Computation*. Volume 6, No. 4, 361 – 386, 1998

[Mic95]   Z. Michalewicz. Heuristic Methods for Evolutionary Computation Techniques. *Journal of Heuristics*. Volume 1, No. 2, 177 – 206, 1995

[MT90]   S. Martello and P. Toth, *Knapsack Problems*, John Wiley & Sons, 1990

[Ols94]   A. L. Olsen. Penalty functions and the knapsack problem. In *Proceedings of the 1st IEEE International Conference on Evolutionary Computation*, 554 – 558, 1994

[Rai98]   G. R. Raidl. An Improved Genetic Algorithm for the Multiconstrained 0-1 Knapsack Problem. In *Proceedings of the 5th IEEE International Conference on Evolutionary Computation*, 207 – 211, 1998

[Rai99]   G. R. Raidl. Weight-Codings in a Genetic Algorithm for the Multiconstraint Knapsack Problem. In *Proceedings of the Congress on Evolutionary Computation*, 596 – 603, 1999

[RPLH89]   J. T. Richardson, M. R. Palmer, G. Liepins, and M. Hilliard. Some Guidelines for Genetic Algorithms with Penalty Functions. In J. D. Schaffer (ed.), *Proceedings of the Third International Conference on Genetic Algorithms*, 191 – 197, Morgan Kaufmann, 1989

[RS95]   G. Rudolph and J. Sprave. A Cellular Genetic Algorithm with Self-Adjusting Acceptance Threshold. In *Proceedings of the 1st IEE/IEEE International Conference on Genetic Algorithms in Engineering Systems: Innovations and Applications*, 365 – 372, IEE, London, 1995

[RS96]   G. Rudolph and J. Sprave. Significance of Locality and Selection Pressure in the Grand Deluge Evolutionary Algorithm. In H.-M. Voigt, W. Ebeling, I. Rechenberg, and H.-P. Schwefel (eds.), *Parallel Problem Solving from Nature - PPSN IV*, 686 – 695, Lecture Notes in Computer Science, Volume 1141, Springer, 1996

[TV94]   J. Thiel and S. Voss. Some Experiences on Solving Multiconstraint Zero-One Knapsack Problems with Genetic Algorithms. *INFOR*, Volume 32, No. 4, 226 – 242, 1994

[Yan98]   R. Yang. *Line-Breeding Schemes for Combinatorial Optimization*, 448 – 457, in [EBSS98]

[ZT98]   E. Zitzler and L. Thiele. *Multiobjective Optimization Using Evolutionary Algorithms – A Comparative Case Study*, 292 – 301, in [EBSS98]

Characterizing Locality in Decoder-Based EAs for the Multidimensional Knapsack Problem

Jens Gottlieb[1] and Günther R. Raidl[2]

[1] Department of Computer Science, Technical University of Clausthal,
Julius-Albert-Str. 4, 38678 Clausthal-Zellerfeld, Germany
`gottlieb@informatik.tu-clausthal.de`
[2] Institute for Computer Graphics, Vienna University of Technology,
Karlsplatz 13/1861, 1040 Vienna, Austria
`raidl@apm.tuwien.ac.at`

Abstract. The performance of decoder-based evolutionary algorithms (EAs) strongly depends on the locality of the used decoder and operators. While many approaches to characterize locality are based on the fitness landscape, we emphasize the explicit relation between genotypes and phenotypes. Statistical measures are demonstrated to reliably predict locality properties of selected decoder-based EAs for the multidimensional knapsack problem. Empirical results indicate that (i) strong locality is a necessary condition for high performance, (ii) the concept of heuristic bias also strongly affects solution quality, and (iii) it is important to maintain population diversity, e.g. by phenotypic duplicate elimination.

1 Introduction

Locality is known as an important factor for well-working evolutionary algorithms (EAs) [3,11,14,22]. Although locality can be interpreted in several ways, all interpretations are motivated by the same basic idea: Small changes in genotype performed by evolutionary operators such as mutation and crossover should result in small changes of phenotype, where phenotype is identified by the represented solution or its fitness. EAs which do not fulfill this condition at least partly act like pure random search, hence are not efficient. Thus, the design process of EAs for any problem should be guided by the locality principle [22], which is sometimes also termed principle of strong causality [21].

Many EAs are based on decoders, which map the genotype of a solution onto its phenotype. Obviously, such approaches can only be successful if the employed decoder supports locality. The goal of this paper is to present a new technique for characterizing locality of decoder-based EAs with particular emphasis on the

C. Fonlupt et al. (Eds.): AE'99, LNCS 1829, pp. 38–52, 2000.

multidimensional knapsack problem (MKP), which is stated as

$$\text{maximize} \quad \sum_{j \in J} p_j x_j \tag{1}$$

$$\text{subject to} \quad \sum_{j \in J} r_{ij} x_j \leq c_i, \quad i \in I \tag{2}$$

$$x_j \in \{0, 1\}, \quad j \in J \tag{3}$$

with $I = \{1, \ldots, m\}$ and $J = \{1, \ldots, n\}$ denoting the sets of resources and items, respectively. The MKP is a prominent example for an NP-complete combinatorial optimization problem with a wide range of applications [4]. Therefore, many exact and heuristic algorithms have been developed for the MKP and diverse variants of it [2,15], and in particular several EAs were proposed [7].

Since it is difficult to analyze even simple EAs using direct encoding and primitive operators, most approaches to characterize locality are based on empirical rather than theoretical investigations. Often, locality is characterized by correlation measures based on fitness landscapes [3,11,14], however approaches based on that idea do not directly consider the effects of the used decoder and operators. Therefore we propose a locality concept which is independent of actual fitness values and explicitly examines the structural effects of decoder and employed mutation and crossover operators. Our approach involves measurements that can be applied without performing time-consuming EA runs. In this way it becomes possible in advance to identify and discard EAs which do not provide strong locality. Since the basic principle of our locality characterization is presented in a general fashion, it can be adapted to other decoder-based EAs.

This paper is organized as follows. Section 2 provides an overview of decoder-based EAs of varying complexity for the MKP together with an empirical comparison of the achieved performance on a standard MKP test suite. Our approach to characterize locality is introduced in Sect. 3. Section 4 presents empirical measurements performed for four selected decoder-based EAs. Our results point at important properties and differences of these EAs, which help to explain the achieved performance. Conclusions are given in Sect. 5.

2   Decoder-Based EAs for the MKP

Many EAs with different constraint handling strategies have been proposed for the MKP, see [7] for a survey. On the one hand, there are approaches directly working in the *phenotyic search space* $P = \{0, 1\}^n$. They rely on penalizing or repairing infeasible solutions. The currently best EAs for the MKP we are aware of [2,7,17] are based on heuristic repair and local improvement methods to focus search on the boundary of the feasible region, which is known to contain the optimum [6]. On the other hand, some EAs proceed by exploring an arbitrary *genotypic search space* G, which is mapped into P by a *decoder*. Such a decoder usually employs problem-specific knowledge ensuring to generate only feasible solutions, hence no penalties or repair methods are necessary to deal with infeasible solutions, and usually simple operators can be used. These *decoder-based EAs* perform an indirect search for solutions, and therefore their success strongly depends on the employed decoder. Obviously, such EAs can only be successful if the fittest parts of P are covered, i.e. there exist genotypes which decode into these phenotypes. Furthermore, a decoder is required to be computationally fast, since otherwise the evaluation of many solution candidates would be too time-consuming. Another important factor is how well a decoder supports locality. Since our goal is to characterize several decoder-based EAs concerning this last aspect, we proceed by a brief introduction and empirical comparison of them.

2.1 Permutation Based EA

The *permutation based EA* (PBEA) has been proposed by Hinterding for the (unidimensional) knapsack problem [9] and can easily be adapted to the MKP [17,24]. A solution candidate is represented by a permutation $\pi : J \to J$ of the items. The decoder starts with the feasible solution $x = (0,\dots,0)$ and traverses all variables x_j in the order determined by π, increasing the corresponding variable from 0 to 1 if this does not violate any resource constraint. Hinterding employs standard permutation operators, namely uniform order based crossover and swap mutation which randomly exchanges two different positions.

2.2 Ordinal Representation Based EA

The *ordinal representation based EA* (OREA) was originally considered for the traveling salesperson problem (TSP) [8], however its application to MKP is straightforward. Solution candidates are represented by a vector v with $v_a \in \{1,\dots,n-a+1\}$ for $a \in J = \{1,\dots,n\}$. The decoder is based on a list initially containing all items in a predefined order and starts with the MKP solution $x = (0,\dots,0)$. Items are iteratively removed from the list and checked for inclusion in the solution. In detail, v is scanned from the first to its last position, interpreting each entry v_a as a position in the current list. Such a position identifies the next item j, for which x_j is increased to 1 if the resource constraints remain satisfied. Since each checked item is removed from the list, its size decreases by 1 during each step and reaches length 1 when the last item is to be selected. This representation has the interesting property that classical one-point crossover is applicable because resulting offsprings always represent legal solutions. Moreover, a simple mutation operator can be used which randomly chooses a position a and then draws v_a from $\{1,\dots,n-a+1\}$. However, a closer look at the decoding procedure reveals that a change in a single position of v might have a major impact on the decoded solution since each item selection modifies the list, thus, influences all following item selections. According to our locality conception we expect OREA to yield bad results due to its weak locality.

2.3 Surrogate Relaxation Based EA

Raidl proposed the *surrogate relaxation based EA* (SREA) which represents so-lution candidates by real-valued weights for the items [18]. These weights are used to temporarily modify the profits p_j in the objective function (1) yielding a similar but slightly different MKP instance. This biased problem is then sol-ved by a surrogate duality based heuristic. The solution obtained in this way is also feasible for the original, unbiased problem since the resource constraints (2) remain unchanged. The heuristic, which has originally been proposed by Pirkul [16], starts with the solution $x = (0, \ldots, 0)$ and traverses all items according to decreasing profit/pseudo-resource consumption ratio. Variables x_j are set to 1 if no resource constraint is violated. Pseudo-resource consumptions are deter-mined via reasonable surrogate multipliers which are obtained from the linear programming (LP) relaxed MKP. Since this process would result in solving the LP relaxation for each solution candidate, Raidl suggests to determine the sur-rogate multipliers only once for the original problem in a preprocessing step to decrease the computational effort [18]. SREA uses uniform crossover and a mutation operator which is applied 3 times to each new genotype, modifying a randomly chosen weight by resetting it to a new random value. The results of SREA are the best among all decoder-based EAs for the MKP we are aware of.

2.4 Lagrangian Relaxation Based EA

The *Lagrangian relaxation based EA* (LREA) was also proposed by Raidl and is basically equivalent to SREA, except for the heuristic used to generate a so-lution for the biased problem [18]. LREA employs the procedure introduced by Magazine and Oguz [13] to obtain a solution via Lagrangian relaxation. Since the determination of exact Lagrange multipliers is too time-consuming, some reasonable (but usually suboptimal) multipliers are calculated by a simpler heu-ristic. Each obtained solution is then locally improved by traversing the variables according to decreasing profit and increasing them if feasibility can be maintai-ned.

2.5 Comparison of Decoder-Based EAs

All considered EAs are based on decoders which use resource information to produce only solutions on the boundary of the feasible region, hence the search concentrates on the most promising parts of the phenotype space. Nevertheless the EAs significantly differ in the employed problem specific knowledge. While the decoders of PBEA and OREA ignore profit information, the heuristics em-ployed by LREA and SREA strongly depend on it. Thus, SREA and LREA exploit more knowledge about the problem structure. We remark another inte-resting relation between the decoders of PBEA, OREA, and SREA: Both OREA and SREA internally produce a permutation of the items which is then inter-preted in exactly the same fashion as PBEA decodes solution candidates.

Table 1. Average results of the EAs

m	n	gap [%]				duplicate ratio [%]			
		PBEA	OREA	SREA	LREA	PBEA	OREA	SREA	LREA
5	100	0.53	1.19	0.53	0.52	6.40	34.92	12.11	5.89
5	250	0.25	2.05	0.17	0.16	4.33	33.34	4.28	2.97
5	500	0.21	3.18	0.07	0.10	4.62	34.00	3.62	2.56
10	100	1.00	1.59	1.00	1.00	5.12	36.14	6.84	3.81
10	250	0.57	2.33	0.34	0.35	4.88	36.14	5.11	2.39
10	500	0.54	3.29	0.19	0.25	4.93	36.62	4.58	2.24
30	100	1.78	2.76	1.67	1.70	7.72	38.76	8.73	3.62
30	250	0.97	3.46	0.75	0.87	6.36	38.27	7.52	2.94
30	500	0.85	4.00	0.47	0.59	7.30	37.98	6.49	2.87
total		0.74	2.65	0.58	0.62	5.74	36.24	6.65	3.27

We compare the described decoder-based EAs on a standard test suite of MKP benchmark problems introduced by Chu [2] and available from the OR-Library[1]. The test suite contains 10 instances for each combination of $m \in \{5, 10, 30\}$, $n \in \{100, 250, 500\}$, and tightness ratio $\alpha \in \{0.25, 0.5, 0.75\}$ (each problem has been generated randomly such that $c_i = \alpha \sum_{j \in J} r_{ij}$ holds for all $i \in I$). We selected the first problem of each category yielding a total of 27 problems and performed three runs for each instance. A similar general setup as in [2,7,17,18] was chosen for all EAs, namely population size 100, parent selection via tournaments of size 2, steady-state replacement (replacing the worst individual), crossover probability 1.0, duplicate elimination (a newly generated individual is only accepted if it is not already contained in the population), and an evaluation limit of 200 000 non-duplicate solutions. We observed that in particular duplicate avoidance is essential to prevent an overcrowding of the population by many duplicates of only few different solutions (premature convergence) [19]. Duplicates should be identified on phenotypic rather than genotypic level, i.e. an individual is rejected if its decoded solution is already represented in the current population. The solution quality is measured by the *gap* of the objective value w.r.t. the optimal value of the LP-relaxed problem, i.e. $1 - max^{\text{EA}}/opt^{\text{LP}}$ with max^{EA} and opt^{LP} denoting the best objective value found by the EA and the optimal value of the LP relaxation of MKP, respectively.

Table 1 shows average results determined from the 9 runs per m, n-combination and EA. The *duplicate ratio* (DR) represents the ratio of rejected duplicates among all generated solutions. As expected, OREA yields the worst gap. Furthermore, the high DR indicates that the used operators tend to produce many duplicates. The other EAs perform quite well compared to OREA, so they should also provide locality. The best quality is obtained by SREA, probably due to the employed heuristic. LREA achieves the lowest DR, however PBEA and SREA also yield an acceptable DR which is an order of magnitude smaller than that of OREA. We conclude that the operators of PBEA, SREA,

---

[1] http://mscmga.ms.ic.ac.uk/info.html

and LREA mostly generate new solutions. In general, we consider PBEA, SREA, and LREA to be well adapted to the MKP, in contrast to OREA which is viewed as an example for a badly designed decoder-based EA. The rest of this paper examines the effects of locality, which helps to explain the results from Table 1.

3   Measures for Locality Characterization

Many approaches were proposed to predict an EA's performance for a given problem. Since several techniques are based on different interpretations of locality, we briefly review them to enable a clear distinction from our new approach to measure the locality of decoder-based EAs. Many proposals from literature are based on the fitness landscape which enables an examination of the relation between solution candidates and their fitness values. Beside some theoretical proofs of convergence rates for EAs applied to relatively simple test functions (e.g. evolution strategies applied to the corridor or the sphere model [20]), Manderick et al. proposed to use correlation measures to examine the effects of operators [14]. They randomly generated parents, applied crossover to produce offsprings, and then calculated the correlation coefficient for the average fitness of parents and offsprings. Their approach predicted the performance of several operators in case of NK-landscapes and the TSP and has also been used for several other problems, e.g. minimum span frequency assignment [25]. Fogel and Ghozeil suggested to focus on the operators' abilities to produce offsprings with higher fitness than the parents [3]. Their model also considers the parent selection strategy to reflect actual EA dynamics and has been used for real-valued problems and the TSP. A different approach termed fitness distance correlation (FDC) was investigated by Jones and Forrest for classical genetic algorithms using binary encoding [11]. FDC is based on the intuition that fitness values should reflect the distance to an optimal solution. They proposed to randomly generate solution candidates and calculate the correlation of their fitness values to the distances to the optimum. For this purpose the Hamming distance is used, but a distance metric relying on the operators which actually define the edges of the landscape graph would be more appropriate. Jones and Forrest reported their approach being a reliable predictor of performance on the examined problems [11], however Altenberg provided a counterexample to show that such approaches might be misleading if actual EA dynamics are not considered [1]. In general, all formerly discussed approaches have some drawbacks limiting their ability to predict performance [1,12], hence alternative approaches should be devised.

While most fitness based techniques only implicitly consider the encoding and used operators, in particular for decoder-based EAs the locality characteristics of these parts seem to be most important, hence should rather be explicitly examined. Locality should alternatively be interpreted in terms of the explicit relation between genotypes and phenotypes, which emphasizes the structural effects of operators and the employed decoder. Sendhoff et al. derived such locality conditions, motivated by the claim that small genotypic changes should imply small phenotypic changes [23]. Their probabilistic measures, which are solely focused on the mutation operator, were successfully applied to continuous parameter optimization and structure optimization. Based on Sendhoff et al.'s approach, Igel examined the probabilistic measures for *NK*-landscapes and problems in the genetic programming context [10]. He concluded that the proposed conditions are helpful to compare several codings and operators for a given problem.

We propose a statistical locality concept which considers crossover and mutation operators, enabling a separate analysis of each.

3.1   Distance Metrics in Genotype and Phenotype Spaces

To characterize the locality of the operators within the genotype space G and the phenotype space P, we quantify the distance of two arbitrary solutions in both G and P. Therefore we introduce distance metrics to measure how many different *properties* (either genotypic or phenotypic) are present in two solutions.

For the MKP, the definition of a *phenotypic distance metric*

$$d_P(x,y) := \sum_{j \in J} |x_j - y_j| \quad \text{for } x, y \in P \tag{4}$$

is straightforward, since the Hamming distance counts the number of variables with different values (different phenotypic properties) in the two solutions. For other combinatorial optimization problems, the phenotypic distance usually needs to be defined in some different, meaningful way, which might not always be as obvious as for the MKP. E.g. in case of the TSP, the total number of different edges might be an appropriate measure since edges can be seen as the most important phenotypic properties of TSP solutions [8].

The definition of a *genotypic distance metric* $d_G(X, Y)$ for $X, Y \in G$ is not that straightforward as it depends on the specific encoding and the operators. To remain general, we implicitly define $d_G(X, Y)$ via the mutation operator:

1. Two identical genotypes have distance 0, i.e. $d_G(X, X) = 0$ for $X \in G$.
2. Two distinct genotypes X, Y have distance $d_G(X, Y) = 1$ and are called *adjacent* if the probability to produce Y from X by a single mutation is greater than 0.
3. In general, a genotypic distance $d_G(X, Y) = k$ means that at least k mutations are necessary to transform X into Y.

For the considered EAs, the proposed definitions of $d_G(X, Y)$ and $d_P(x, y)$ satisfy the metric conditions, namely identity, symmetry, and the triangular inequality.

3.2   Mutation Innovation *MI*

Usually, the mutation operator is not applied exactly once but either with a given probability or a certain number of times. Let $X^m \in G$ represent the solution obtained from $X \in G$ by applying mutation with exactly this probability or rate. Note that according to the previous definitions, k consecutive mutations will produce an offspring with $d_G(X, X^m) \leq k$.

Let x and x^m be the phenotypes corresponding to X and X^m. Assuming X to be a random variable with uniform distribution within G, X^m, x, and x^m are dependent random variables. We define the *mutation innovation* as

$$MI := d_P(x, x^m),\qquad(5)$$

which describes how much phenotypic "innovation" is introduced by the mutation. MI is a random variable, whose distribution immediately reflects several important aspects concerning locality of mutation.

We have to consider the case $MI = 0$, occurring with probability $P(MI = 0)$ and meaning that the mutation has not affected the phenotypic properties. Large values for $P(MI = 0)$ indicate that either mutation often does not change any genotypic properties or many different genotypes are mapped to the same phenotype, which reflects a high degree of redundancy in G. There are two possible reasons for such high redundancy.

Firstly, $|G|$ might be significantly larger than $|P|$. Often such a representational redundancy decreases performance, but sometimes it may also be beneficial and lead to better final results [22].

A second reason may be that the decoder contains local improvement techniques or heuristics which always or mostly lead to preferred phenotypes in a restricted subset $P' \subset P$. We call this effect *heuristic bias*. Therefore, solutions $x \in P \setminus P'$ cannot be represented or have substantially smaller probabilities to be generated. While such a restriction of P might sometimes be advantageous, it must be ensured that promising areas and particularly the global optima are covered [18,22]. As already mentioned, the four considered EAs for the MKP restrict the search space to the boundary of the feasible region, therefore, they work with heuristic bias.

$P(MI = 0)$ can principally be controlled by tuning the mutation probability or rate. Obviously, a higher mutation rate would decrease $P(MI = 0)$, but the usually resulting larger changes in genotype may also affect the EA to behave more like inefficient random search. Now, consider only the cases in which mutation produces an offspring x^m which actually differs from x. Then the expected value for MI under this restriction, called $E(MI|MI > 0)$, and the standard deviation $\sigma(MI|MI > 0)$ are good indicators for the locality of mutation. Only if both $E(MI|MI > 0)$ and $\sigma(MI|MI > 0)$ are reasonably small, successful mutations lead in general to similar phenotypes. Large values signalize that very different solutions are frequently generated, only negligible or none locality is given, and hence the search of the EA tends to be a random search.

3.3 Crossover Innovation CI_k

When using binary crossover, a new genotype $X^c \in G$ is generated from two parental solutions $X^{p_1}, X^{p_2} \in G$. Let $x^{p_1}, x^{p_2}, x^c \in P$ be the phenotypes corresponding to X^{p_1}, X^{p_2}, X^c. Usually, the result of the crossover operation is strongly influenced by the similarity of the parents. In early stages of an EA run the population has high diversity, hence most selected parents differ significantly, while in later stages the population is likely to be converged, i.e. similar parents are frequently involved in crossover applications. To reflect population diversity in an approximate way using our notion of genotypic distance, let us assume that X^{p_2} is produced by applying $k \geq 1$ consecutive mutations to X^{p_1}, i.e. the genotypic distance between the parents is $d_G(X^{p_1}, X^{p_2}) \leq k$. Additionally, we take the duplicate elimination of the EAs into account by considering only phenotypicly non-identical parents, therefore we presume $d_P(x^{p_1}, x^{p_2}) > 0$. We then define the *crossover innovation*

$$CI_k := \min(d_P(x^c, x^{p_1}), d_P(x^c, x^{p_2})) \tag{6}$$

as the phenotypic distance of x^c to its closer parent. If we interpret the genotypes X^{p_1}, X^{p_2} as random variables with the restriction $d_G(X^{p_1}, X^{p_2}) \leq k$, then $X^c, x^{p_1}, x^{p_2}, x^c$, and in particular CI_k are dependent random variables.

Obviously, CI_k is 0 if either $x^c = x^{p_1}$ or $x^c = x^{p_2}$. Analogously to $P(MI = 0)$, the probability $P(CI_k = 0)$ indicates the likelihood for crossover generating an offspring which is phenotypicly identical to one of its parents. Usually, $P(CI_k = 0)$ is higher for parents with very similar or equal genotypes, i.e. for small k. A high $P(CI_k = 0)$ for large k indicates that crossover does not mix genotypes well or the degree of redundancy in the mapping $G \rightarrow P$ is high. Especially when both probabilities $P(MI = 0)$ and $P(CI_k = 0)$ are high, the EA cannot work efficiently since many duplicate solutions are generated.

Analogously to the mutation, we restrict our considerations in the following to the case of crossover actually producing new solutions with $CI_k > 0$. The expectations $E(CI_k | CI_k > 0)$ and standard deviations $\sigma(CI_k | CI_k > 0)$ for different maximum parent distances k are meaningful indicators for the existence or absence of locality during crossover: For strong locality, $E(CI_k | CI_k > 0)$ should be small for small k and become increasingly larger for larger k. In particular, large values for both, $E(CI_k | CI_k > 0)$ and $\sigma(CI_k | CI_k > 0)$, for small k are strong indicators for weak locality.

3.4   Crossover Loss CL_k

Besides the ability to generate new solutions with adequate distances to their parents, an important aspect of crossover is that a generated solution mainly consists of properties inherited from its parents; only few new properties should be introduced. The importance of such behavior is e.g. described by the building-block hypothesis [5]. To consider this aspect, too, we define the *crossover loss* CL_k as the number of phenotypic properties of x^c which are not inherited from either x^{p_1} or x^{p_2} but are newly introduced. For the MKP this means

$$CL_k := \sum_{j \in J} \delta(x_j^c, x_j^{p_1}, x_j^{p_2}) \tag{7}$$

$$\text{with} \quad \delta(x_j^c, x_j^{p_1}, x_j^{p_2}) = \begin{cases} 0 & \text{if } x_j^c = x_j^{p_1} \text{ or } x_j^c = x_j^{p_2} \\ 1 & \text{otherwise.} \end{cases} \tag{8}$$

Table 2. Empirically estimated characteristics for MI

measure	PBEA	OREA	SREA	LREA
$P(MI = 0)$ [%]:	49.06	49.04	65.22	14.43
$E(MI|MI > 0)$:	5.34	27.87	4.47	6.16
$\sigma(MI|MI > 0)$:	1.55	22.93	1.44	2.72

Considering also the proposed phenotypic distance metric, the crossover loss can alternatively be written as

$$CL_k := \frac{1}{2}(d_P(x^c, x^{p_1}) + d_P(x^c, x^{p_2}) - d_P(x^{p_1}, x^{p_2})). \qquad (9)$$

Clearly, $CI_k = 0$ implies $CL_k = 0$. To prevent a bias by the case where crossover does not produce a new solution, we preclude that case and consider the expected values $E(CL_k|CI_k > 0)$ for different maximum parent distances k. Large values for any k immediately signalize weak locality.

4   Empirical Measurements

Determining the different measures introduced in Sect. 3 for specific encoding techniques and crossover and mutation operators in a theoretical way is in general a very hard task and especially for more complex EAs like SREA or LREA practically nearly impossible. Therefore, we use an efficient empirical approach to obtain good estimations for specific problem instances. Since these measurements can be performed a priori to any EA run, inadequate encoding techniques or operators providing weak locality can be early recognized.

The empirical results presented in this section are based on an average sized MKP instance with $m = 10$, $n = 250$, and $\alpha = 0.5$, namely the first problem of Chu's benchmarks with these parameters [2]. Although different absolute values have been obtained as results for other instances, this specific problem is representative in the sense that the same basic tendencies have been observed for other instances, too.

MI has been empirically estimated by randomly drawing 20 000 genotypes $X \in G$ and applying mutation to each with the same probability or rate as in the EA. Figure 1 shows histograms for the distributions of MI gained in this way, while Table 2 subsumes the estimations for the measures $P(MI = 0)$, $E(MI|MI > 0)$, and $\sigma(MI|MI > 0)$ introduced in Sect. 3.2.

Most noticeable are the generally large values for $P(MI = 0)$. One reason for this effect is that all four EAs generate only phenotypes on the boundary of the feasible region, therefore in a substantially restricted phenotype subset $P' \subset P$. Furthermore, there exists an encoding redundancy in all EAs due to the different sizes of G and P. While $|P| = 2^n$, the genotype space has size $|G| = n!$ in PBEA and OREA and is even larger for SREA and LREA since real values are allowed for each weight. Although the used mutation rate for both SREA and LREA was

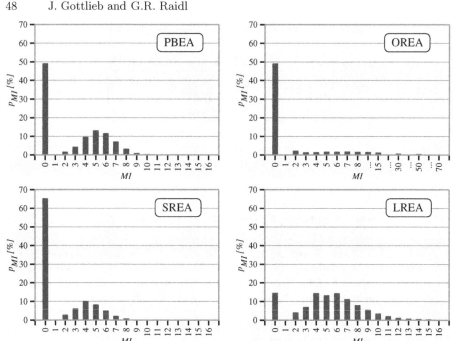

Fig. 1. Empirically determined distributions of MI for the different EAs

3 (instead of only 1 as in PBEA and OREA), SREA has the highest probability $P(MI = 0)$. The reason is the small biasing factor[2] γ of SREA, which means that the original problem is in general significantly less biased than in LREA [18]. Therefore, the heuristic bias of SREA is clearly stronger than that of LREA. Note that $P(MI = 1) = 0$ for all EAs since $d_P(x, y) \geq 2$ for two distinct solutions $x, y \in P$ lying on the boundary of the feasible region. Regarding $E(MI|MI > 0)$ and $\sigma(MI|MI > 0)$, OREA yields substantially higher values than the other EAs. This result is an immediate indicator for the weak locality of the mutation operator and encoding in OREA and can be explained by the strong dependency of each gene's interpretation from all its predecessors in the genotype. According to the other values for $E(MI|MI > 0)$ and $\sigma(MI|MI > 0)$, SREA provides the highest locality followed by PBEA and LREA.

CI_k and CL_k were empirically estimated for $k \in \{2^i \mid i = 0, \ldots, 9\}$ by randomly generating 20 000 parents X^{p_1} for each k, applying k mutations to each X^{p_1} to obtain the associated second parent X^{p_2}, and then producing offsprings X^c via crossover. According to Sect. 3.3, parents representing identical phenotypes $(d_P(x^{p_1}, x^{p_2}) = 0)$ were discarded. Obtained estimations for $P(CI_k|CI_k = 0)$, $E(CI_k|CI_k > 0)$, $\sigma(CI_k|CI_k > 0)$, and $E(CL_k|CI_k > 0)$ are shown in Fig. 2.

The curves for $P(CI_k = 0)$ indicate that PBEA has in general the lowest probability for generating a child identical to one of its parents. Especially for $k = 1$, no other EA is able to produce a solution different from its parents. While

---

[2] Biasing factors were set as proposed in [18]: LREA: $\gamma = 0.2$, SREA: $\gamma = 0.05$.

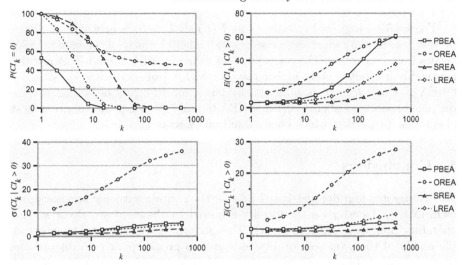

Fig. 2. Empirically determined estimations for $P(CI_k = 0)$, $E(CI_k|CI_k > 0)$, $\sigma(CI_k|CI_k > 0)$, and $E(CL_k|CI_k > 0)$

$P(CI_k = 0)$ decreases with increasing k down to 0 for PBEA, SREA, and LREA, $P(CI_k = 0)$ remains above 45% for OREA. The reason is that OREA uses one-point crossover, which might frequently exchange genes having no effect on the decoded phenotype, because the phenotypic properties are mainly determined by the first genes. Note that the large duplicate ratio of OREA during actual EA runs (see Sect. 2.5) could have been predicted by the high probabilities $P(CI_k = 0)$ and $P(MI = 0)$. Especially for SREA but also for LREA, k must be relatively high to mostly obtain new solutions that are different from their parents. This observation emphasizes the importance of taking care of the population diversity in the EA by discarding generated phenotypic duplicates in order to enable crossover to work efficiently (and hence prevent premature convergence).

For small k, the expected values $E(CI_k|CI_k > 0)$ are reasonably small and nearly equal for PBEA, LREA, and SREA, but relatively high for OREA. This immediately indicates relatively strong locality for crossover of PBEA, LREA, and SREA, but weak locality for OREA.[3] With increasing k, $E(CI_k|CI_k > 0)$ becomes larger for all EAs. The final values which are quite different for the four EAs indicate that PBEA and OREA are in general capable of generating more innovative solutions than LREA and especially SREA. Since the uniform crossover operator of SREA and LREA mixes genotypes at least as well as the crossover operators in the other EAs, the reason for the smaller CI_k for larger k is again the higher heuristic bias towards a smaller phenotype subset P'. The weak locality of OREA is also clearly indicated by the large standard deviations $\sigma(CI_k|CI_k > 0)$ compared to the other EAs' corresponding values.

---

[3] Note that $E(CI_1|CI_1 > 0)$, $\sigma(CI_1|CI_1 > 0)$, and $E(CL_1|CI_1 > 0)$ are not defined for OREA, LREA, and SREA since CI_1 is always 0.

The smallest expected crossover loss $E(CL_k|CI_k > 0)$ is achieved by SREA (always less than 3) which indicates that x^c inherits nearly all phenotypic properties from its parents. The corresponding values for PBEA and LREA are also reasonably low but slightly higher than for SREA (for increasing k), while for OREA $E(CL_k|CI_k > 0)$ is substantially larger for all values of k. This reflects the poor capabilities of crossover in OREA to build offsprings by inheriting most phenotypic properties, which once again implies weak locality.

5   Conclusions

We investigated four decoder-based EAs for the multidimensional knapsack problem (MKP), focusing on the locality achieved by the employed decoders and operators (mutation and crossover). The performed experiments demonstrate the ability of the proposed locality measures to predict poor performance due to weak locality. In case of the MKP, SREA, which is the best performing EA according to the results in Sect. 2.5, offers also the strongest locality regarding mutation and crossover. This is clearly indicated by the small estimated values for $E(MI|MI = 0)$, $E(CI_k|CI_k > 0)$ (for small k), associated standard deviations, and $E(CL_k|CI_k > 0)$. The weak locality of OREA, which is proved by the substantially larger estimations for these measures, is one reason why this EA performs significantly worse than the other considered EAs.

Another important aspect of all four EAs is signalized by the generally high probabilities $P(MI = 0)$ and $P(CI_k = 0)$ for small k: Due to heuristic bias and/or redundancy in the coding, the proportion of operator applications which actually lead to new, different phenotypes may be considerably small. It is therefore important to actively maintain enough diversity in the population, e.g. by discarding generated phenotypic duplicates.

The prediction capabilities of the a priori measurements have also been verified by determining the proposed measures online during actual EA runs (details were not presented here). The online results differed only slightly from the a priori results, provided that a proper k, which can be derived from the population diversity during the EA run, is used for the comparisons regarding CI_k and CL_k.

Generally, our results confirmed locality to be a necessary condition for decoder-based EAs to work well for MKP. However, we are aware that locality is not sufficient for good performance. Also the concept of heuristic bias strongly affects the achieved performance, thus should be examined more detailled. Although the presented locality measures can be used as indicators, it is interesting to check whether additional measures could also reliably predict heuristic bias and hence the total performance. The proposed ideas should also be validated on different problems to verify whether our results can be generalized to hold for other problems, too.

References

1. L. Altenberg: *Fitness Distance Correlation Analysis: An Instructive Counterexample*, in Proc. of the 7th Int. Conf. on Genetic Algorithms, East Lansing, MI, pp. 57 – 64, 1997
2. P. C. Chu, J. E. Beasley: *A Genetic Algorithm for the Multidimensional Knapsack Problem*, Journal of Heuristics 4, pp. 63 – 86, 1998
3. D. B. Fogel, A. Ghozeil: *Using Fitness Distributions to Design More Efficient Evolutionary Algorithms*, in Proc. of the 3rd IEEE Int. Conf. on Evolutionary Computation, Nagoya, Japan, pp. 11 – 19, 1996
4. M. D. Garey, D. S. Johnson: *Computers and Intractability: A Guide to the Theory of NP-Completeness*, Freeman, San Francisco, 1979
5. D. E. Goldberg: *Genetic Algorithms in Search, Optimization and Machine Learning*, Addison-Wesley, 1989
6. J. Gottlieb: *Evolutionary Algorithms for Multidimensional Knapsack Problems: The Relevance of the Boundary of the Feasible Region*, in Proc. of the Genetic and Evolutionary Computation Conf., Orlando, FL, p. 787, 1999
7. J. Gottlieb: *On the Effectivity of Evolutionary Algorithms for Multidimensional Knapsack Problems*, in Proc. of Artificial Evolution, Dunkerque, France, 1999
8. J. J. Grefenstette, R. Gopal, B. Rosmaita, D. Van Gucht: *Genetic Algorithms for the Traveling Salesman Problem*, in Proc. of the 1st Int. Conf. on Genetic Algorithms, Hillsdale, NJ, pp. 160 – 168, 1985
9. R. Hinterding: *Mapping, Order-independent Genes and the Knapsack Problem*, in Proc. of the 1st IEEE Int. Conference on Evolutionary Computation, Orlando, FL, pp. 13 – 17, 1994
10. C. Igel: *Causality of Hierarchical Variable Length Representations*, in Proc. of the 5th IEEE Int. Conf. on Evolutionary Computation, Anchorage, AL, pp. 324 – 329, 1998
11. T. Jones, S. Forrest: *Fitness Distance Correlation as a Measure of Problem Difficulty for Genetic Algorithms*, in Proc. of the 6th Int. Conf. on Genetic Algorithms, Pittsburgh, PA, pp. 184 – 192, 1995
12. L. Kallel, M. Schoenauer: *A Priori Comparison of Binary Crossover Operators: No Universal Statistical Measure, but a Set of Hints*, in Proc. of 3rd European Conf. on Artificial Evolution, Nîmes, France, pp. 287 – 299, 1997
13. M. J. Magazine, O. Oguz: *A Heuristic Algorithm for the Multidimensional Zero–One Knapsack Problem*, European Journal of Op. Res., 16, pp. 319 – 326, 1984
14. B. Manderick, M. de Weger, P. Spiessens: *The Genetic Algorithm and the Structure of the Fitness Landscape*, in Proc. of the 4th Int. Conf. on Genetic Algorithms, pp. 143 – 150, 1991
15. S. Martello, P. Toth: *Knapsack Problems: Algorithms and Computer Implementations*, J. Wiley & Sons, 1990
16. H. Pirkul: *A Heuristic Solution Procedure for the Multiconstrained Zero-One Knapsack Problem*, Naval Research Logistics 34, pp. 161 – 172, 1987
17. G. R. Raidl: *An Improved Genetic Algorithm for the Multiconstrained 0–1 Knapsack Problem*, in Proc. of the IEEE Int. Conf. on Evolutionary Computation, Anchorage, AL, pp. 207 – 211, 1998
18. G. R. Raidl: *Weight-Codings in a Genetic Algorithm for the Multiconstraint Knapsack Problem*, in Proc. of the Congress on Evolutionary Computation, Washington DC, pp. 596 – 603, 1999.

19. G. R. Raidl, J. Gottlieb: *On the Importance of Phenotypic Duplicate Elimination in Decoder-Based Evolutionary Algorithms*, in Late-Breaking Papers Proc. of the Genetic and Evolutionary Computation Conf., Orlando, FL, pp. 204 – 211, 1999.
20. I. Rechenberg: *Evolutionsstrategie: Optimierung technischer Systeme nach Prinzipien der biologischen Evolution*, Frommann-Holzboog, Stuttgart, 1973
21. I. Rechenberg: *Evolutionsstrategie '94*, Frommann-Holzboog, 1994
22. S. Ronald: *Robust Encodings in Genetic Algorithms*, in D. Dasgupta, Z. Michalewicz (eds.), Evolutionary Algorithms in Engineering Applications, pp. 29 – 44, Springer, 1997
23. B. Sendhoff, M. Kreutz, W. von Seelen: *A condition for the genotype-phenotype mapping: Causality*, in Proc. of the 7th Int. Conf. on Genetic Algorithms, East Lansing, MI, pp. 73 – 80, 1997
24. J. Thiel, S. Voss: *Some Experiences on Solving Multiconstraint Zero-One Knapsack Problems with Genetic Algorithms*, INFOR 32, pp. 226 – 242, 1994
25. C. Valenzuela, S. Hurley, D. Smith: *A Permutation Based Genetic Algorithm for Minimum Span Frequency Assignment*, in Proc. of the 5th Int. Conf. on Parallel Problem Solving from Nature, Amsterdam, The Netherlands, pp. 907 – 916, 1998

Evolutionary Case-Based Design

Mike Rosenman

Key Centre of Design Computing
Department of Architectural and Design Science
University of Sydney
NSW 2006 Australia
Email: mike@arch.usyd.edu.au

Abstract. The paper provides a solution to the current problem of adaptation in case-based design. It presents a case-based model of design, using an evolutionary approach, for the adaptation of previously stored design solutions. It is argued that such a knowledge-lean methodology is more general in its applicability than conventional case-based design and that a fundamental genotype representation allows for using the case base in a variety of domain and problem situations. A prototypical example in the context of 2-D spatial design of houses is used to test the efficacy and efficiency of this approach.

1   Introduction

Case-based design, (CBD) [1-4], is an important area in the computability of design. While there has been much work in the indexation and retrieval of cases [5], many issues remain to be resolved. In particular, the adaptation process has, to date, stymied the utility of the approach, as each problem area has demanded problem-specific adaptation knowledge which has proven to be very difficult to formulate.

Evolutionary algorithms [10-13] cover several approaches but in general the evolutionary approach is a general knowledge-lean methodology aimed at producing solutions to difficult and complex problems where the relationships between the solution and the requirements is not well known. Within its methodology is the mechanism for adapting members to a particular given problem. An evolutionary approach to the process of adaptation provides a general problem-independent methodology to the process of case-based adaptation. A general representation of design cases based on a genotypic representation of designs can be used with the general evolutionary process of reproduction and stochastic selection serving as the adaptation process. Moreover, by working with the entire population of existing design solutions, it makes the indexation and selection of relevant cases redundant. In addition, the general genotypic representation allows for multiple phenotypic representations enabling the case base to be used for different problem solving situations. Another major advantage of an evolutionary approach is the capability to provide a number of potentially satisfactory solutions, rather than a single solution, from which a designer can make a final selection.

C. Fonlupt et al. (Eds.): AE'99, LNCS 1829, pp. 53-72, 2000.
© Springer-Verlag Berlin Heidelberg 2000

Evolutionary case-based design (ECBD), will be exemplified through the 2-D spatial design of houses using a genotypic representation of hierarchical form growth developed previously [6-9] for the description of test cases. The outcome is the generation of new and diverse satisfactory (house) design solutions, under new design situations.

2   An Evolutionary Approach to Design

Evolutionary techniques have lately been used in design research [14-16], both for optimization of parameters and for the generation of designs and exploration of design spaces [17-18, 6-9, 19-24]. Recently, there has been some work on evolutionary methods in case-based design [25-29].

2.1   A Hierarchical Model of Evolutionary Synthesis of Form

A hierarchical genotype representation for the generation of form is used [6-9]. In a multi-level evolutionary approach, at each level, a component is generated from a combination of components from the level immediately below. At the element level, an element is generated from a combination of fundamental cells. Thus the building blocks at each level are the components generated at the level immediately below and at each level the genotype contains genes which are assemblies of lower level genes. Through this approach, only those factors relevant to the design of a particular component or assembly need be considered for that component or assembly. These building blocks or components are labeled by a single symbol replacing the string containing the lower level genes in much the same way as automatically defined functions (ADFs) are represented in genetic programming, [30] but the representation used here is much more strictly hierarchical in nature. A particular ADF can be used at any level whereas in this representation there is a strict component/assembly hierarchical order. In addition, each component is evaluated according to its own particular requirements rather than by an overall single evaluation function acting over the whole structure.

Genotype and Phenotype. A design grammar, based on the method for conjoining counteractive vectors in the construction of polygonal shapes and counteractive faces in the construction of polyhedral shapes is used. Counteractive vectors are vectors which are equal and opposite and counteractive faces are faces which have edge sequences which are equal and opposite. A polygon (polymino for orthogonal geometries) may be described by its sequence of edge vectors which is a phenotype description whereas a polyhedron may be described by a sequence of faces and vectors. A gene consists of two components (shapes) and their respective counteractive vectors or faces. The genotype, for a complex shape, is the sequence of such genes. The following will give a simple description in the domain of polyminoes. For more complex and detailed descriptions including those for polyhedra see Rosenman [31,32]. The phenotype representation of a (unit) square using an edge vector representation is (W1, N1, E1, S1,) where the symbols W, N, E, S represent the vector types and the subscripts the instances of that vector type. The counteractive edge vectors are N:S and E:W. A polymino can be grown by adding one unit square in turn to the current shape at any edge. Two polyminoes can be joined through the conjoining of any pair of counteractive edges. There are several possible joinings possible in both growing a polymino and in joining two polyminoes. Figure 1 shows a) a unit square; b) a domino with its genotype and phenotype descriptions and c) the result of joining two polyminoes through the conjoining of the counteractive edges, E2 and W2. The same result could be achieved by joining counteractive edges E1 and W3 or N2 and S1 or S1 and N1 or E3 and W1.

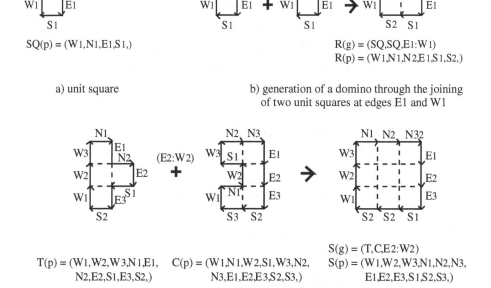

a) unit square

$SQ(p) = (W1,N1,E1,S1,)$

b) generation of a domino through the joining of two unit squares at edges E1 and W1

$R(g) = (SQ,SQ,E1:W1)$
$R(p) = (W1,N1,N2,E1,S1,S2,)$

$T(p) = (W1,W2,W3,N1,E1,$ $N2,E2,S1,E3,S2,)$

$C(p) = (W1,N1,W2,S1,W3,N2,$ $N3,E1,E2,E3,S2,S3,)$

$S(g) = (T,C,E2:W2)$
$S(p) = (W1,W2,W3,N1,N2,N3,$ $E1,E2,E3,S1,S2,S3,)$

c) joining of polyminoes, T and C, at E2 and W2 to form a third polymino, S

Fig 1. Growing and joining polyminoes

Note that, while in the above description, the phenotype interpretation of the genotype is a sequence of edge vectors representing a description of a shape, and is useful for perimeter information, the phenotype interpretation could differ for different problems. For example, a representation based on the centre coordinates of the cells can be found from the genotype description and can be used when the area, centre of gravity and second moment of area of a shape are required.

Hierarchical Evolution In The Domain Of House Plans. The domain of house plans will be used to describe the hierarchical nature of the representation although the representation and concepts are general in nature. A house can be composed of zones where zones are composed of rooms and rooms, in turn, composed of a number of unit space units, unit squares, depending on the area of the room required. In a hierarchical evolutionary process, a room is generated by firstly randomly generating a number of shapes of the given number of unit squares and then evolving the shapes over a number of generations. A number of satisfactory shapes can be selected for each room. Each zone is the generated by initially generating a population of zone solutions composed of one of each required room type randomly chosen from each room type set generated and through randomly selected edge joinings. The zone population is then evolved to arrive at a number of satisfactory zone solutions. The house is generated and evolved in a manner similar to that of the zones, by randomly selecting and joining zones and evolving the population. Figure 2 shows the house hierarchy.

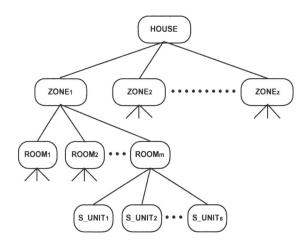

Fig2. House hierarchy

Figure 3 shows the general and hierarchical evolutionary process. Figure 3 is diagrammatic only as the hierarchy may occur over any number of component/assembly levels and evolution occurs over many generations. The final composition of components and objects is thus a mix of several elements from those contained at the initial generation even though it appears as directly composed from the initial random generation. The pseudocode for this hierarchical evolutionary process is as follows:

```
For each level of the component hierarchy, l = 1, L
    For each component type at level l, ct = 1, N
        Generate a random population Po of m members for ct by assembling
            lower level building blocks, one component from each component type at level l-1
        Evolve population Po through a number of runs each of a number of
            generations. At each generation select and save suitable components
        Stop when have selected sufficient satisfactory components for ct
        Selected components for ct become building blocks for level l+1
    Stop when all component types at level l evolved
Stop when all levels of hierarchy done, i.e. have reached top level or design level
```

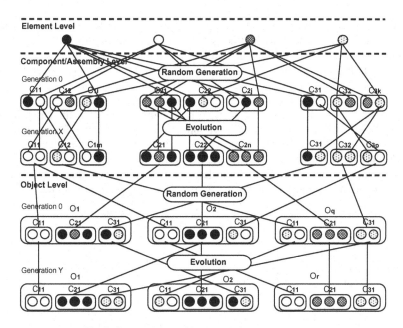

Fig 3. General hierarchical evolutionary process

Examples of genotypes and phenotypes for each level of the house hierarchy are shown in Figure 4, where the phenotypic information is the shape of the component. In the genotype description of rooms, the unit square has been omitted for simplicity and the genotype represented as a list of edge vector conjoinings. In the description of the phenotype, multiple instances of the same edge type are indicated as G1,...n where G indicates the edge type (N ,E, S or W) and n the last instance of that type in the sequence, so that W1,...3, indicates W1, W2, W3.

Element	Shape	Genotype	Phenotype
Room DR1		(E1:W1,N2:S1,N1:S1,W2:E1, S1:N1,W3:E1,N3:S1,E3:W1)	(N1,E1,E2,E3,S1,S2,S3,W1,W2,W3,N2,N3,)
Zone LZ1		((LR2,DR1,E4:W2),EN1,N3:S1)	(N1,2,3,E1,...,5,S1,...7,W1,...5,N4,...7,)
House H1		(LZ1, BZ1, N7:S5)	(W1,...5, N1,...7, E1,..5,S1,E6,...,10, S2,...,8,W6,...10,)

Fig 4. Example of room, zone and house genotype and phenotype

Simple crossover is used for the propagation of members during the evolution process.The roulette wheel method is used for selection of members for 'mating' and for selection of members for new generations. Although the examples use 2-D spatial forms for houses, using a square cell, the methodology is more general and the grammar can be applied to any polygonal or polyhedral configurations.

Figure 5 shows the interface to the system as well as the results of the house design. The interface to the system uses three windows. The main window is the DRAWING WINDOW where the populations of rooms, zones and houses appear for each generation. When a user selects a solution it appears in the CURRENT SELECTIONS window. In this example 12 houses have been selected so far from previous runs and this run. The bottom-left window shows the graph monitoring the best value and the average value for the fitness over the generations so far.

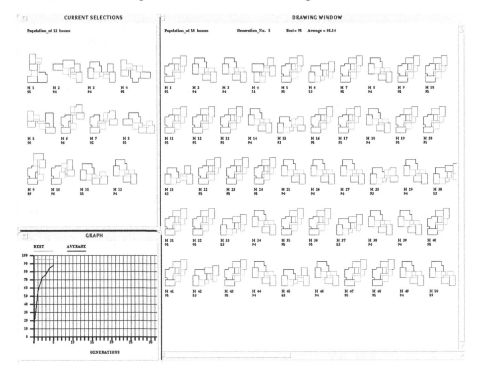

Fig5. Interface of generative system showing results of house generation

3   Conventional Case-Based Design - CCBD

Conventional Case-Based design [1-4] is derived from the general method of case-based reasoning, or CBR [33,34]. According to Schank [33], the most relevant experience is selected and 'tweaked' accordingly to produce the required solution to the given problem.

CCBD is based on reusing specific design experiences rather than using general process knowledge. A number of specific design experiences or cases are stored in a case base. When a new problem is given, rather than using process knowledge, such as rules, to generate a solution, the closest-matching case is found in the case base and domain knowledge then applied to adapt it to the new problem The main problems faced by CCBD are:

1. *the representation of design cases*: it is still a matter of discussion as to what exactly should be in a case, i.e. what its contents, structure and representation should be [35]. The description of a case is predetermined according to a particular view of the type of information expected to be relevant.

2. *indexing*: the ability to access information in a case base depends on the indexing schema used which itself depends on the nature of and representation used for the case contents. Thus, only the information which is explicitly represented can be used for case indexing and hence accessed. This prescribes the nature of the queries and hence type of problems that can be addressed.

3. *matching*: the matching of existing cases to the new problem is done in a number of ways such as counting the number of attribute entries in the case which match those of the required problem. So that finding matched cases depends on the predetermined information structure and contents.

4. *retrieval and selection*: two strategies are a) to find only a 'best-match' case and return it, in which instance there is no further selection; and b) to select several 'close-match' cases and return these to the user for final selection. In both instances, there needs to be an evaluation for finding 'best-match' or 'close-match'.

5. *adaptation*: While much work has been applied to the first four problems and various techniques exist [3,4], the problem of adaptation still remains a major obstacle to the general application of the approach. There is no general knowledge on how to 'tweak' especially in the design domain and each design application and problem needs its specific adaptation knowledge. In addition, there is no guarantee that a so-called 'best-match' or 'close-match' will be amenable to adaptation in an efficient way. The selected case may be so tightly-coupled that any modification leads to disintegration of the design. To date there has been little if any research on the potentiality of adaptation when selecting a case for adaptation.

So, in general, the limitations of CCBD are the difficulties encountered in the adaptation process and that, while design problems are situated in given problem situations which demand different information, the explicit fixed representation of case bases, in terms of the attributes described, prescribe the type of problems that can be addressed.

4   Evolutionary Case-Based Design - ECBD

CCBD is based on Schank's model of CBR [33] whereas evolutionary case-based design (ECBD), in contrast, uses Calvin's model [36-38]. In Schank's model a particular closest match experience is selected and 'tweaked' (adapted) whereas in alvin's model thinking tasks are achieved by Darwinian competition to organize complex patterns.

Existing house designs can be formulated using the above hierarchical representation resulting in a case base of house designs represented by their genotype which encodes how they are to be synthesized. Given a particular set of requirements formulated as a fitness function, this case base is used in its entirety as the initial population for an evolutionary process which will evolve new and satisfactory solutions to the given problem. Starting from a population of existing house designs rather than generating a solution from the beginning has the advantage that higher level components, rooms and zones, already exist. An example of this is given in Figure 6 where two house solutions, H1 and H2, with low perimeter to area ratios, are selected as parents to produce, after crossover, the solutions H3 and H4. The solutions found are satisfactory solutions to a design problem using a fitness function of high perimeter to area ratio which reflects the need for solutions for cross ventilation in a hot humid climate.

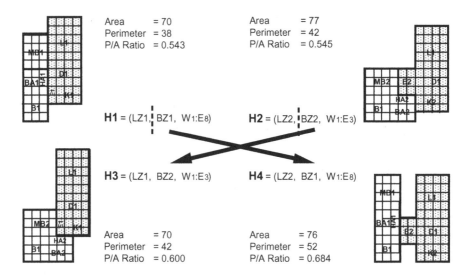

Fig 6. Evolution of house plans resulting from a new fitness function

One of the shortcomings in conventional case-based design, CCBD, is the concentration on selecting the best-matched case in the assumption that this will lead to simple adaptation. This assumption breaks down when the selected case is so tightly constrained that any 'tweaking' causes major perturbances. It may well be that a solution not so closely matched is actually easier to adapt. In an evolutionary case-based design approach, ECBD, it is not necessary to preselect members for adaptation as the general methodology of the approach will progressively remove less fit members from future populations. Moreover since direct relationships between the structure (form) and function (performance) are not always known, seemingly poor performing solutions may have elements which are useful in combination with other elements of other members. Figure 7 shows a diagrammatic comparison between the CCBD and ECBD approaches.

Figure 7 shows the different modules corresponding to the different functions required, the necessity for CCBD to have a knowledge base for adaptation knowledge, containing both domain specific and general knowledge and the fact that only one solution is produced with CCBD as against several solutions in ECBD. The domain specific knowledge, together with the case representation, prescribes the type of problems that can be solved. All possible situations must be accounted for. In contrast, the evolutionary process has a general adaptation methodology. Domain specific knowledge is built into the fitness function and hence into the interpretations of the genotype and evaluation of the phenotypes.

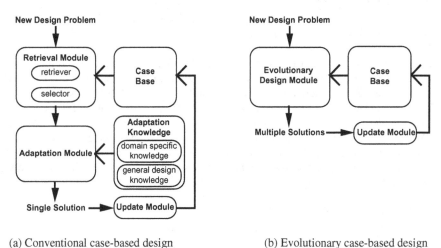

(a) Conventional case-based design   (b) Evolutionary case-based design

Fig 7. Diagrammatic comparison between conventional and evolutionary case-based design

5   Relation to Existing Work on Evolutionary Case-Based Design

There is existing work in the area of evolutionary case-based design by Gero and Schnier [25], Maher and Gomez da Silva Garza [26] and   Louis [27]. Gero and Schnier use more direct learning in a genetic engineering approach to first evolve useful building blocks (complex genes) and then reuse them in combination with the basic genes in a manner similar to GP. The learning is carried out from a set of examples containing both good and bad solutions. The work presented here uses a more structured representation where the components are meaningful entities subject to their own requirements rather than just substrings. Moreover, there is no requirement of the case base containing any good solutions although obviously there must be sufficient possibility to evolve to good solutions. While both Louis and Maher and Gomez have similar conceptual approaches, both carry out preselections on the case base before carrying out evolution. In the case of Maher and Gomez this results in a very small population for the evolution process. Louis mentions that a problem exists in deciding how to select the population so as not to lose some potentially critical information. Moreover, Louis's example is in the domain of scheduling rather than design synthesis. Maher and Gomez use a flat genotype representation where genes directly represent attributes of buildings. Thus the representation in this work, which stems from the concept of hierarchical aggregation of atomic cells into more complex elements, as well as the use of the entire case base forms a major difference between this work and other work.

6   Implementation

6.1   Case Base

A case base of 40 houses was formulated using the genotypic representation previously described. Each house consists of 3 zones, namely the Living Zone, Utility Zone and Bed Zone. There are 10 Living Zone types, 12 Bed Zone types and 2 Utility Zone types. Each zone type can be rotated 90, 180 or $270°$, (R90, R180 or R270) or mirrored along the X or Y axes (MX or MY) or mirrored and rotated (MXR90, MYR90). The genotype of each house is given as the zones selected and edges of the zones joined. Thus each house genotype is $3 + 2$ (zones + edge joinings) long. The genotypes of each zone and room is not given here but as an example, the following zone genotypes are:

Element	Genotype
LZ1	(((LR1, DR1, W6:E1), Ki1, S7:N1), En1, E7:W4)
UZ1	La1
BZ2	(((MB2, Ha2, N5:S1), B1, N1:S4), Ba1, E1:W2), B1(R90), E3:W2)

So that the genotype of house, H1, can be given in a tree form as

(((((LR1, DR1, W6: E1), Ki1, S7:N1), En1, E7:W4), (La1(R90)), S1:N3),
((((MB2, Ha2, N5:S1), B1, N1:S4), Ba1, E1:W2), B1(R90), E3:W2), S3:N5)

as shown in Figure 8. The room genotypes as generations of unit squares are not given here, for simplicity, but could be similarly expanded and added to the tree representation. Figure 9 shows the 40 cases used.

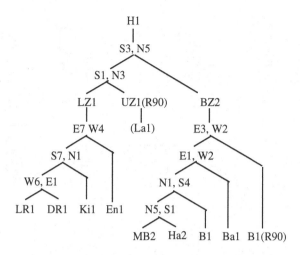

Figure 8. Tree representation of house H1 genotype

6.2 New Problem Formulation

As a first test to investigate possible issues, a design problem was formulated where the aim is to design houses suitable for hot humid climates, i.e. requiring the maximum cross-ventilation possible. As an approximation, this was formulated as maximizing the perimeter to area ratio, P/A of a house. In addition it was required that the circulation efficiency of the house remained reasonably efficient. That is, that the required relationship between the rooms in the house be satisfactory. The problem could have been formulated as a Pareto optimization problem with two criteria, namely maximizing the P/A ratio and minimizing a weighted sum of distances between rooms with given closeness requirements. However, it was felt that house with very bad circulation efficiencies (CE) might very well produce large P/A ratios but be totally unsuitable. Hence the fitness function was formulated to include a penalty function to penalize solutions for poor circulation efficiency. The fitness function was then formulated as:

$$\text{Max } F = (P/A * 100) * P$$

where P is a penalty and

P = 1	if CE <= 100
P = 0.75	if 100 < CE <= 150
P = 0.5	if 150 < CE <= 250
P = 0.25	if 250 < CE <= 500
P = 0.1	if 500 < CE

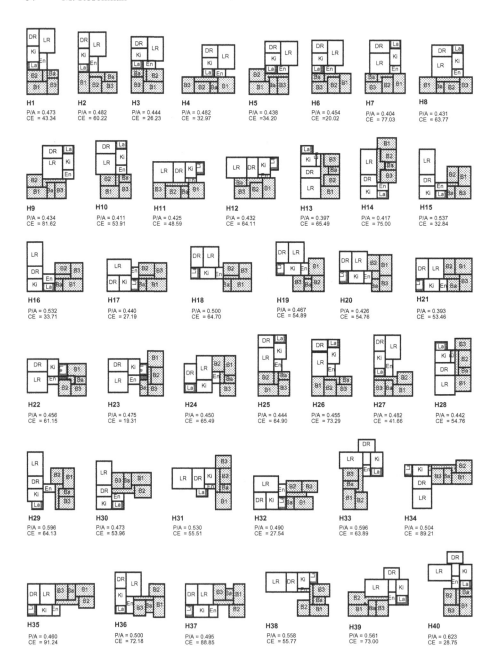

Fig9. Test case base of 40 houses

The penalty values were obtained from inspection of house solutions with acceptable and near acceptable circulation efficiencies. In Figure 9, the fitness of each house is given in terms of its perimeter to area (P/A) as well as its circulation efficiency (CE). The cases are divided into two groups, cases 1 to 26 which are in general very ompact houses and cases 27 to 40 which include some less compact houses, the aim being to test what information is necessary for a case base to produce satisfactory results. For houses 1 to 26 the lowest P/A is 0.393 (house H21), the highest is 0.537 (house H15) and the average P/A is 0.450. For houses 27 to 40 the lowest P/A is 0.442 (house H28), the highest is 0.624 (house H40) and the average P/A is 0.523, giving an average P/A of 0.475 for the 40 cases.

6.3   Tests

A total of 5 tests were carried out. Each test was carried out over a number of runs with each run over a number of generations until 10 solutions deemed satisfactory were found. The first four tests are case-based tests while the fifth test is a generation of houses using the growth process to be used as a comparison against the case-based results. This growth process involves the generation of rooms, zones and houses. The tests are as follows:

- Test 1: crossover limited to the house level, i.e. only between zones and the joining edges, i.e. only the top part of the tree in Figure 8 was allowed for cut sites. So that only the zones with the orientation and the edge joinings as existing in the cases could be used in new combinations - 26 cases.
- Test 2: crossover limited to the house level as for Test 1but allowing random edge joinings between zones – 26 cases
- Test 3: as for Test 1 – 40 cases
- Test 4: as for Test 2 – 40 cases
- Test 5: growth process

For the case base of 26 cases, there are 10 Living Zone types, 12 Bed Zone types and 2 Utility Zone types. The augmentation to 40 cases resulted in 11 Living Zone types, 14 Bed Zone types and 2 Utility Zone types. For Test 5, 16 Living Zone types, 2 Utility Zone types and 12 Bed Zone Types were generated.

The same interface was used and satisfactory designs selected as previously noted.

6.4   Results

Figures 10, 11, 12, 13 and 14 show the 10 house results found for Tests 1 to 5 respectively.

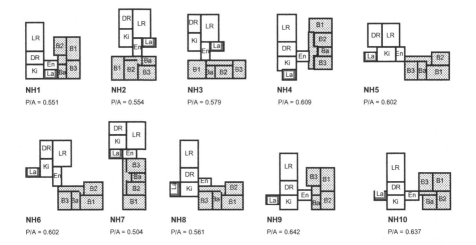

Figure 10. Ten houses resulting from Test 1 – 26 cases

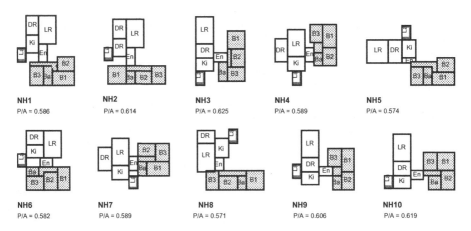

Figure 11. Ten houses resulting from Test 2 – 26 cases

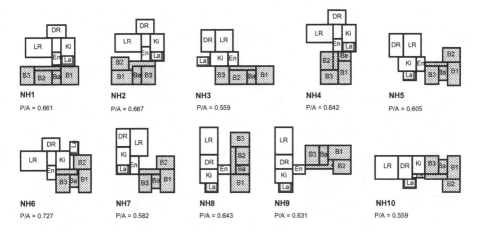

Figure 12. Ten houses resulting from Test 3 – 40 cases

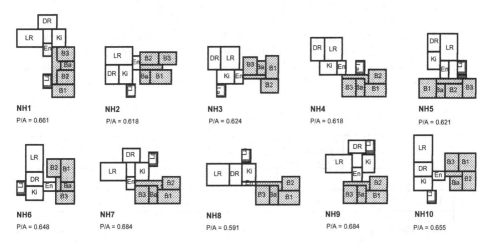

Figure 13. Ten houses resulting from Test 4 – 40 cases

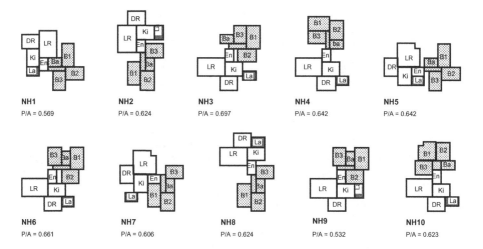

Figure 14. Ten houses resulting from Test 5 – growth process

Table1 shows some of the statistics from the five tests. The number of runs and generations for Test 5 are the total for all rooms, zones and the houses. The number of runs for the rooms and zones was 36 and the number of generations for the rooms and zones was 567. Thus, for Test 5 the number of generations for the houses was 290.

Table 1. Statistics

	No. of Cases	Runs	Total No. of Generations	Max. No. of Generations in a Run	Best P/A	Average P/A
Test 1	26	15	217	30	0.642	0.584
Test 2	26	19	162	27	0.625	0.596
Test 3	40	7	70	13	0.727	0.628
Test 4	40	21	170	17	0.684	0.640
Test 5	-	63	857	60	0.697	0.622

6.5   Analysis of Results

The two tests using the 26 very compact houses, i.e. Tests 1 and 2 obtained fairly similar results. Although the maximum P/A result of 0.642 for Test 1 (solution 1-NH9) was higher than the maximum of 0.625 for Test 3 (solution 2-NH3) the average P/A result for Test 2 of 0.596 was higher than the average of 0.584 of Test 1. A similar result is obtained when comparing Tests 3 and 4. The maximum P/A of 0.727 for Test 3 (solution 3- NH6) is higher than the maximum of 0.684 for Test 4 (solutions 4-NH7 and 4-NH9) whereas the average of 0.640 for Test 4 was higher than the average of 0.628 for Test3. This shows that allowing the zones to be randomly joined does not necessarily result in obtaining a solution that is better than using the joinings given in the case base but does result in getting a better set of solutions from which a final choice can be made. Since the aim in design is not necessarily to find a solution with the 'best' based on a narrow definition of 'best' given by some fitness function but may be based also on other factors not expressed in the fitness function, getting a range of potentially good solutions is important.

On the other hand there is a marked difference between the results of the tests using 26 cases (Tests 1 and 2) and the results of the test using 40 cases (Test 3 and 4). It is quite clear that expanding the case base to 40 cases improves both the maximum P/A and average P/A results and that the number of generations required to obtain the results drops. This shows that the efficiency of the case-based approach depends strongly on the information available in the case base. If the genotypic information is poor with respect to its capability of producing fit solutions then unless mutation can change this the results will be impoverished. When there exists sufficient genotypic information then the case-based approach is very efficient.

There is not a great difference in the quality of results obtained from the case-based approach using 40 cases and the results obtained by generating the houses anew. However, the effort expended by the case-based approach is substantially less. For Test 2, a total of 70 generations were run whereas for Test 5 a total of 857 generations were run (counting the generations of rooms, zones and houses). This shows that the information contained in the case base regarding the rooms, composition of zones and the joining of zones is useful.

A feature which appeared in the generation of houses anew (Test 5) and when allowing random joining of zones (Tests 2 and 4) was that the initial random population generated was very uneven. Only very few (1 in most cases) solutions had values which were close to being satisfactory, the rest had very low values because of the poor circulation efficiency values resulting from poor joinings of zones. So that while one or two solutions may have a score of approximately 40 (P/A x Penalty), many solutions had score of less than 10. This meant that one or two solutions clearly dominated the rest of the population and the run quickly converged to those solutions. In some cases, where all the initially generated solutions were very low, the run converged without any satisfactory solutions found. This explains why the number of generations with Tests 2 and 4 is fairly high. In comparison, the initial generation for Tests 1 and 3 was the case base with a fairly even spread of solutions.

7   Summary and Future Work

The evolutionary case-based design process promises a general, domain independent, method for case-based design with solutions to the problems of case-base adaptation, case representation and indexing. The experimentation has shown that even at the highest building block level, it is possible to achieve good results. In addition, not only is it unnecessary to preselect 'good' cases or classify the cases into 'good' and 'bad' solutions but this may be detrimental. This contradicts the conclusions reached by Louis and Johnson [29] and Maher and Gomez de Silva Garza [26]. That is, all the information that is available in the design experience, represented by the case base, should be used since it is not possible to determine what components of any cases may or may not be useful by simply evaluating the performance of the overall design.

However, what is critical is the information contained in the case base. This is not specific to the evolutionary approach but to case-based design in general. A case base must contain sufficient information with the potential to generate suitable solutions from its cases to be effectual.

Future work is needed to resolve several issues, such as:

- how large and how varied does the case-base need be to be useful?
- the level at which the genetic operations, e.g. crossover, have to be applied in a hierarchical representation. Can they be restricted to the highest level of description for the house, i.e. using zones as building blocks or will it be necessary to descend to lower levels, e.g. to create new zones or even new rooms? This is allowing cut sites at different levels in the tree structure. Unlike the general genetic programming process the crossover in the tree representation has to be controlled between similar parts, or subtrees, of the tree to maintain consistency of structure.
- related to the previous point, what information can be learnt from a case base and reapplied and what new information needs be created? Should only the component information be used or both the component and configuration (joinings) of component information be used? For example, it may be that only the zone (and room) solutions can be reused, without the need for new solutions, whereas the rotation and edge joinings have to be generated anew.
- since the efficiency of a hierarchical case base is related to the reuse of high-level components, if new lower-level components are to be generated, a strategy is required to bias the process towards reuse rather than regeneration.

Further, the use of evolutionary case-based design brings into focus the interpretation of the genotype into phenotype descriptions depending on the particular problem. Since in the test example, the fitness function was associated with perimeter values, an interpretation based on edge vectors was suitable. However, other problems may require other phenotypic interpretations. This addresses the issue that although the process for creating the solution is given by the genotype description, its description is not fixed a priori but determined depending on the problem.

Acknowledgements

This work is supported by Australian Research Council Large Grants, A8970053

References

1.   Flemming U. (1994). Case-based design in the SEED system, *in* G. Carrara and Y. Kalay, (eds), *Knowledge-Based Computer-Aided Architectural Design*, Elsevier, Amsterdam, Holland, pp.69-91.
2.   Oxman R. and Oxman R. (1994). Case-based design: cognitive models for case libraries, *in* G. Carrara and Y. Kalay, (eds), *Knowledge-Based Computer-Aided Architectural Design*, Elsevier, Amsterdam, Holland, pp.45-68.

3. Maher M. L., Balachandran M., and Zhang D. M. (1995). *Case-Based Reasoning in Design*, Lawrence Erlbaum, Hillsdale, NJ.
4. Maher M. L., and Pu P (eds). (1997). *Issues and Applications of Case-Based Reasoning in Design*, Lawrence Erlbaum, Hillsdale, NJ.
5. Williams P. (1995), Dynamic Memory for Design, *PhD Thesis*, (unpublished), Department of Architectural and Design Science, University of Sydney, Sydney Australia.
6. Rosenman M. A. (1996a). A growth model for form generation using a hierarchical evolutionary approach, *Microcomputers in Civil Engineering*, Special Issue on Evolutionary Systems in Design, **11**:161-172.
7. Rosenman M. A. (1996b). The generation of form using an evolutionary approach, *in* J. S Gero and F. Sudweeks (eds), *Artificial Intelligence '96*, Kluwer Academic, Dordrecht, Germany, pp.643-662.
8. Rosenman M. A. (1997a), The generation of form using an evolutionary approach, *in* D. Dasgupta and Z. Michalewicz (eds), *Evolutionary Algorithms in Engineering Applications*, Springer-Verlag, Southhampton and Berlin, pp.69-85.
9. Rosenman M. A. and Gero J. S. (1999). Evolving designs by generating useful complex gene structures, *in* P. J. Bentley (ed.), *Evolutionary Design by Computers*, Morgan Kaufmann, San Fransisco, pp.345-364.
10. Holland J. H. (1975). *Adaptation in Natural and Artificial Systems*, The University of Michigan Press, Ann Arbor.
11. Goldberg D. E. (1989). *Genetic Algorithms in Search, Optimization, and Machine Learning*, Addison-Wesley, Reading, MA.
12. Koza J. (1992). *Genetic Programming: On Programming Computers by Means of Natural Selection*, MIT Press, Cambridge, Mass.
13. Schwefel H-P. (1995). *Evolution and Optimum Seeking*, John Wiley and Sons, New York.
14. Woodbury R. F. (1993). A genetic approach to creative design, *in* J. S. Gero and M. L. Maher, (eds), *Modeling Creativity and Knowledge-Based Creative Design*, Lawrence Erlbaum, Hillsdale, NJ, pp.211-232.
15. Louis S. J. and Rawlins G. J. (1991). Designer genetic algorithms: genetic algorithms in structure design, *in* R. K. Belew and L. B. Booker (eds), *Proc. Fourth Int. Conf. on Genetic Algorithms,* Morgan Kaufmann, San Mateo, CA, pp.53-60.
16. Michalewicz Z., Dasgupta D., Le Riche R. G. and Schoenauer M. (1996). Evolutionary algorithms for constrained engineering problems, *Computers and Industrial Engineering Journal*, Special Issue on Genetic Algorithms and Industrial Engineering, **30**(4): 851-870.
17. Bentley P. J. and Wakefield J. P. (1995). The table: an illustration of evolutionary design using genetic algorithms, *Genetic Algorithms in Engineering Systems: Innovations and Applications*, GALESIA '95, pp.412-418.
18. Bentley P. J. and Wakefield J. P. (1997). Generic Evolutionary Design, *in* P. K. Chawdry, R. Roy and R. K. Pant (eds), *Soft Computing in Engineering Design and Manufacturing,* Springer Verlag, London, pp.289-298.
19. Koza J., Bennett III, F. H., Andre D. and Keane M. A. (1996). Automated designs of both the topology and sizing of analog electrical circuits using genetic programming, *in* J. S Gero and F. Sudweeks (eds), *Artificial Intelligence '96*, Kluwer Academic, Dordrecht, Germany, pp.151-170.
20. Roy R., Parmee I.C. and Purchase G. (1996). Integrating the genetic algorithm with the preliminary design of gas turbine cooling systems, *in* I. C. Parmee (ed.), *Proceedings of 2nd International Conference on Adaptive Computing in Engineering Design and Control*, PEDC, University of Plymouth, 1996.
21. Gero J. S., Kazakov V. and Schnier T. (1997). Genetic engineering and design problems, *in* D. Dasgupta and Z. Michalewicz (eds), *Evolutionary Algorithms in Engineering Applications*, Springer Verlag, Southhampton and Berlin, pp.47-68.

22. Dasgupta D. and Michalewicz Z. (eds). (1997). *Evolutionary Algorithms in Engineering Applications*, Springer-Verlag, Southhampton and Berlin.
23. Bentley P. J. (ed.) (1999. *Evolutionary Design by Computers*, Morgan Kaufmann, San Fransisco.
24. Parmee I. C. (1999). Exploring the design potential of evolutionary search, exploration and optimisation, *in* P. Bentley (ed.), *Evolutionary Design by Computers*, Academic Press (in press).
25. Gero J. S. and Schnier T. (1995). Evolving representations of design cases and their use in creative design, *in* J. S. Gero, M. L. Maher and F. Sudweeks (eds), *Preprints Computational Models of Creative Design*, Key Centre of Design Computing, University of Sydney, pp.343-368.
26. Maher M. L. and Gomez de Silva Garza A. (1996). The adaptation of structural system design using genetic algorithms, *Proceedings of the International Conference on Information Technology in Civil and Structural Engineering - Taking Stock and Future Directions*, Glasgow, Scotland.
27. Louis S. (1997). Working from blueprints: evolutionary learning for design, *Artificial Intelligence in Engineering*, **11**:335-341.
28. Rosenman M. A. (1997b). An exploration into evolutionary models for non-routine design, *Artificial Intelligence in Engineering*, **11**:287-293.
29. Louis, S and Johnson, J. (1999). Robustness of case-initialized genetic algorithms, *Proceedings of FLAIRS-99*, FLAIRS, May 1999, pp.129 - 133.
30. Koza J. (1994). *Genetic Programming II: Automatic Discovery of Reusable Programs,* MIT Press, Cambridge, Mass.
31. Rosenman M. A. (1995). An edge vector representation for the construction of two-dimensional shapes, Environment and Planning B:Planning and Design, **22**:191-212.
32. Rosenman M. A. (1999). A face vector representation for the construction of polyhedra, *Environment and Planning B:Planning and Design*, **26**:265-280.
33. Schank R. C. (1982). *Dynamic Memory: A Theory of Reminding and Learning in Computers and People*, Cambridge University Press, Cambridge, UK.
34. Riesbeck C. K. and Schank, R. C. (1989). *Inside Case-Based Reasoning*, Lawrence Erlbaum, Hillsdale, NJ.
35. Rosenman M. A., Gero J. S. and Oxman, R. E. (1992). What's in a case: the use of case bases, knowledge bases and databases in design, *in* G. N. Schmitt (ed.), *CAAD Futures '91*, Viewig, Wiesbaden, Germany, pp.285-300.
36. Calvin W. H. (1987). The brain as a Darwin machine, *Nature*, **330**:33-34.
37. Calvin W. H. (1996). *How brains think.: evolving intelligence, then and now*, Basic Books, New York.
38. Calvin, W. H. (1998). Competing for consciousness, *Journal of Consciousness Studies*, **5**(4):388-404.

Shorter Fitness Preserving Genetic Programs

Anikó Ekárt

Computer and Automation Research Institute
Hungarian Academy of Sciences
1518 Budapest, POB. 63, Hungary
ekart@sztaki.hu

Abstract. In the paper a method that moderates code growth in genetic programming is presented. The addressed problem is symbolic regression. A special mutation operator is used for the simplification of programs. If every individual program in each generation is simplified, then the performance of the genetic programming system is slightly worsened. But if simplification is applied as a mutation operator, more compact solutions of the same or better accuracy can be obtained.

1 Introduction

An important problem with genetic programming systems is that in the course of evolution the size of individual programs is continuously growing. The programs contain more and more non-functional code. When evaluating the genetic programs over the fitness cases, much of the time is spent on these irrelevant code fragments. Thus, they reduce speed and, in the meantime, make the programs unintelligible for humans.

Koza [7] proposes the use of a maximum permitted size for the evolved genetic programs as parameter of genetic programming systems. Accordingly, genetic programs are allowed to grow until they reach a predefined size. In the same work a so-called editing operation is proposed for (1) making the output of genetic programming more readable and (2) producing simplified output or improving the overall performance of genetic programming. The editing operation consists in the recursive application of a set of simplifying rules. However, it was shown that for the boolean 6-multiplexer problem the application of the editing operation does not influence the performance of genetic programming in a notable way.

Hooper and Flann [5] apply expression simplification in a simple symbolic regression problem. They conclude that the accuracy of genetic programming could be improved by simplification. Additionally, simplification could (1) prevent code growth and (2) introduce new useful constants.

Langdon [8] introduces two special crossover operators, the so-called *size* fair and *h*omologous crossovers. These operators create an offspring by replacing a subtree of one parent with a carefully selected similar-size subtree of the other parent. By using these operators, code growth is considerably reduced without affecting the performance of genetic programming.

C. Fonlupt et al. (Eds.): AE'99, LNCS 1829, pp. 73–83, 2000.
© Springer-Verlag Berlin Heidelberg 2000

There are several studies that suggest taking into account the program size when computing the fitness value.

Iba, de Garis and Sato [6] define a fitness function based on a *Minimum Description Length* (MDL) principle. The structure of the tree representing the genetic program is reflected in its fitness value:

$$mdl = Error\_Coding\_Length + Tree\_Coding\_Length.$$

Zhang and Mühlenbein [16] demonstrate the connection between accuracy and complexity in genetic programming by means of statistical methods. They use a fitness function based on the MDL principle:

$$Fitness_i(g) = E_i(g) + \alpha(g)C_i(g),$$

where $E_i(g)$ and $C_i(g)$ stand for the error and the complexity of individual i in generation g. The Occam factor $\alpha(g)$ is computed as a function of the least error in the previous generation $E_{best}(g-1)$ and the estimated best program size for the current generation $\hat{C}_{best}(g)$. Thus, their fitness function is adaptively changing from generation to generation.

Soule, Foster and Dickinson [12] compare two methods for reducing code growth in a robot guidance problem: (1) the straightforward editing out of irrelevant and redundant parts of code and (2) the use of a fitness function that penalizes longer programs. They conclude that applying the penalty outperforms *any kind* of editing out, so providing new evidence for [6,16].

Notwithstanding, other studies show that these seemingly irrelevant or redundant parts of code are useful because they shield the highly-fit building blocks of programs from the destructive effects of crossover.

Angeline [1] calls these apparently useless fragments of code *introns*, in analogy with the introns contained in DNA. He points out that the formation of introns should not be hindered, since they provide a better chance for the transfer of complete subtrees during crossover.

Banzhaf et al. [2] argue that the analogy to biological introns might be wrong. But since *intron* is already a common term in the genetic programming domain, we will use it throughout the paper with the meaning of *non-functional code*.

Nordin, Francone and Banzhaf [9] demonstrate through experiments that introns allow a population to keep the highly-fit building blocks and in the meantime make possible the protection of individuals against destructive crossover. They introduce the so-called *Explicitly Defined Introns* that are inserted in the code and serve as a control mechanism for the crossover probability of the neighboring nodes.

In the present paper we describe a method that takes both advice into account:

- Code growth in genetic programming should be limited in order to obtain a comprehensible solution in a reasonable amount of time.
- Introns should be preserved, since they shield the highly-fit building blocks from the harmful effects of crossover.

The paper is organized as follows: In Section 2 the biological evidence that inspired this work is presented. In Section 3 the method is described and in Section 4 the results of 800 runs of the system on two symbolic regression problems are shown. Then a real-world application using these results is discussed and conclusions are drawn.

2 Biological Background

The DNA of bacteria contains continuous coding sequences. For many years it was believed that the genes of higher organisms are also continuous. This view was changed in 1977, when it was discovered that the genes of some eukaryotic organisms (made of nucleated cells) are discontinuous, and the non-coding sequences are much longer than the coding sequences.

Generally, the genes of living organisms consist of:

- exons - base sequences that encode proteins or polypeptides; and
- introns - base sequences that do not participate directly in the production of proteins.

The introns can influence the amount and the quality of the proteins expressed by the gene in which they occur. But the mechanism of the indirect effect of introns on protein production is not known. They represent regions in which DNA can break and recombine without affecting the encoded proteins [13].

There is evidence suggesting that: *"introns were present in ancestral genes and were lost in the evolution of organisms that have become optimized for very rapid growth, such as eubacteria and yeast"* [13].

Gilbert [4] points out that *"introns have been used to assemble those genes that are the late product of evolution"*. At the same time, he brings evidence for the *loss* of introns during evolution.

Thus, introns play an important role in evolution, when shielding the exons from destruction through crossover. But they can disappear in the course of evolution.

On the other hand, the main source of variability is mutation. From the many existing types of mutation, we consider the following [2,14,15]:

- point mutation - change of one base pair to another;
- neutral mutation - a genetic change that is neither advantageous nor disadvantageous for the organism;
- frameshift mutation - insertion or deletion of one or more base pairs; and
- large DNA sequence rearrangement.

Usually, in genetic programming systems the analog of the first type is implemented. In addition, we also consider here the other mutation types.

3 The Proposed Method

The goal of this work is to reduce the size of genetic programs evolved in symbolic regression problems. However, the method could be applied to any genetic programming system after defining the corresponding rules that simplify the structure of programs.

We designed a special mutation operator that modifies only the structure of a genetic program; the interpretation and the fitness value remain the same (therefore, it could be seen as the analog of the biological neutral mutation). This modification is intended to eliminate the occasional introns and simplify the structure of the genetic program, without altering its accuracy (in a similar way to the editing operation of [7] and the expression simplifier of [5]). Since the problem is symbolic regression, this mutation operator performs the algebraic simplification of the expression of a genetic program. The simplifier is implemented in Prolog and consists of approximately 250 clauses. Some of the simplification rules are shown in Table 1.

Table 1. Some simplification rules

Original expression	Simplified expression	Binding
$0 + x$	x	
$K_1 * x + K_2 * x$	$K * x$	$K = K_1 + K_2$
$K_1 * K_2 * x$	$K * x$	$K = K_1 * K_2$
$(-1) * x$	$-x$	

Let us see two short examples:

$$f(x, y) = x * 2 - x * (3 - 1) + 3 * x/y \stackrel{\text{simpl}}{=} 3 * x/y$$
$$g(x) = x * (x - 2) + 3 * x \stackrel{\text{simpl}}{=} x + x * x$$

In the first example the non-functional part was removed and in the second one algebraic simplification was performed.

One can see the analogy of this simplification to the biological mutation:

− simplification as a whole - neutral mutation;
− removal of non-functional code - frameshift mutation; and
− algebraic simplification - large DNA sequence rearrangement.

The removal of non-functional code is more restricted than frameshift mutation, since it is applied only to non-functional code and there is no addition, just deletion of this code.

The algebraic simplification is in fact more than a rearrangement. For a more precise analogy with biology we could have made just a simple transformation, such that:

$$g(x) = x * (x - 2) + 3 * x \stackrel{\text{transf}}{=} (3 * x - 2 * x) + x * x.$$

This transformation is closer to large DNA sequence rearrangement and could be the subject of later experiment.

Since the simplification of an algebraic expression involves the removal of non-functional code, we decided for a single "new" mutation operator, that performs all possible simplifications on the selected expression (like the editing operation proposed by Koza [7]).

We applied this mutation operator in addition to the usual recombination operators (crossover and point mutation). We thought that applying the simplification in every generation might be too drastic and time-consuming and, therefore, we made the frequency of its application a parameter of the genetic programming system (also suggested by Koza [7]).

4   Experimental Results

Experiments were conducted on two symbolic regression problems. The goal was to evolve the programs that approximate the functions (1) $F_1(x) = x + x^2 + x^3 + x^4$ and (2) $F_2(x) = 1.5 + 24.3x^2 - 15x^3 + 3.2x^4$, respectively, in 100 data points, that were randomly selected in the $[0, 1]$ interval. The parameter setting is shown in Table 2.

Table 2. The genetic programming parameter setting

Objective	Evolve a function that fits the data points of the fitness cases
Terminal set	x, real numbers $\in [-100, 100]$
Function set	$+, *, /$
Fitness cases	$N = 100$ randomly selected data points (x_i, y_i), (1) $y_i = x_i + x_i^2 + x_i^3 + x_i^4$ (2) $y_i = 1.5 + 24.3x_i^2 - 15x_i^3 + 3.2x_i^4$
Raw fitness and also standardized fitness	$\sqrt{\frac{1}{N}\sum_{i=1}^{N}(gp(x_i) - y_i)^2}$, $gp(x_i)$ being the ouput of the genetic program for input x_i
Population size	100
Crossover probability	90%
Probability of Point Mutation	10%
Selection method	Tournament selection, size 10
Termination criterion	none
Maximum number of generations	50
Maximum depth of tree after crossover	20
Initialization method	Grow
Frequency of simplification	Every 1., 2. or 5. generation
Simplification probability	0-100%

In the fitness measure only the error is included, there is no term related to the program size. We think that *a shorter program with more errors should not be preferred to a longer program containing less errors*. The mechanism for limiting code growth should be distinct from the selection mechanism. We introduced the simplification as a mutation operator in order to reduce code size without influencing the fitness-based selection mechanism.

Table 3. Parameter setting for simplification

Frequency F[gen.]	Probability P[%]
-	0
1	10, 20, 100
2	20, 40, 60
5	20, 25, 50, 75, 80, 100

We added two parameters: the frequency of simplification (F) and the simplification probability (P). Simplification is applied every F-th generation, on every individual program with probability P.

For each parameter setting shown in Table 3 we performed 50 runs and recorded their average. The plots for the regression of F_2 are presented, since it is a more difficult problem. Nonetheless, the plots for F_1 have the same character.

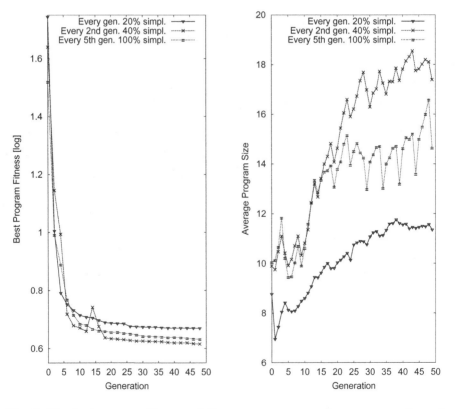

Fig. 1. The best program fitness and the average program size over generations, for $P/F = 10\%$

In order to establish a good ratio between the probability and the frequency of simplification, we compared the results of the runs with the same overall simplification ratio P/F. ($P/F = 10\%$ is achieved when (1) 10% probability of simplification in each generation, (2) 20% probability of simplification in every second generation or (3) 50% probability of simplification in every fifth generation is applied). Particularly, we analyzed the results for $P/F = 10\%$ and $P/F = 20\%$. The results for $P/F = 20\%$ are shown in Figure 1.

Considering program size, best results were obtained when simplification was applied in every generation: code growth practically stopped at the 35th generation. In the case of 100% simplification in every fifth generation, the form of the graphics (Figure 1 right) clearly reflects the alternating behavior of program size: after every fifth generation the average program size is reduced by simplification, then programs are allowed to grow for the next five generations. In fact, if we look at the average program size at every fifth generation (after simplification), we can see that in this case code growth is moderated after the 30th generation. In the meantime, the fitness of the best program (Figure 1 left) was slightly lower when simplification was applied less frequently (and with higher probability, keeping P/F constant).

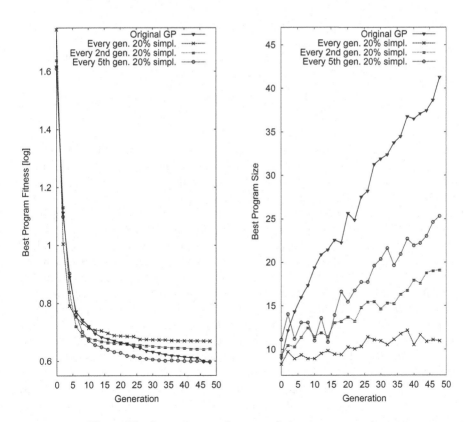

Fig. 2. The best program fitness and size over generations

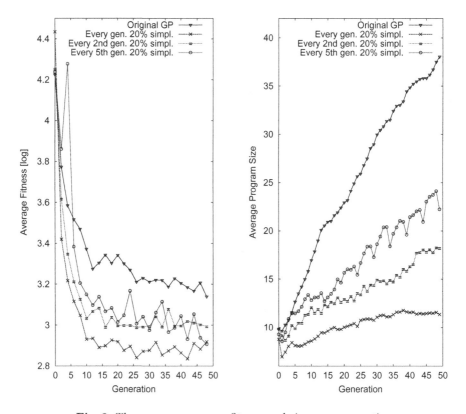

Fig. 3. The average program fitness and size over generations

We made another comparison among the results of runs with different frequencies of simplification, keeping the probability constant at 20%. Figures 2 and 3 show the results for the cases: (1) no simplification (original genetic programming), (2) simplification in every generation, (3) every second or (4) every fifth generation. The results for simplifying every individual in every generation ($F = 1$, $P = 100\%$) are very close to those obtained when simplifying in every generation with probability $P = 20\%$ (since the corresponding graphics are practically undistinguishable, we do not show the graphics corresponding to total simplification). As we expected, in the case of genetic programming without simplification, the program size is continuously growing (on the average, 0.5 *nodes/generation*) and in the case of simplification in every generation, the program size remains at a low value. In the latter case, by averaging over 50 independent experiments, the code growth is 0.08 *nodes/generation* (for the average individual) for the first 28 generations, and 0.02 *nodes/generation* for the last 22 generations. The fitness of the best program (Figure 2) is slightly worse when simplification is applied more often. The average fitness (Figure 3) of the cases when using simplification is better than that of the case with no simplification.

We also compared the results of runs, when the frequency of simplification was constant (every fifth generation) and the probability of simplification varied between 0-100%. While the size of programs was growing fast when no simplification was applied, it was quite stable when the probability of simplification was high. In the case of simplifying each individual program, the accuracy of the best program was slightly worse than for the other cases. Thus, we found again that applying simplification more often leads to much shorter, but slightly worse solutions.

Since a better performing program is a program with lower fitness and lower size, we could represent the performance of genetic programs as the product of fitness and size (Figure 4). In this view, the results are better when simplification is applied more often. But if we represent the performance as $Fitness^{N_F} * Size^{N_S}$, N_F and N_S being integers, if $N_F > N_S$ (meaning that fitness is more important than size), then the differences shown in Figure 4 diminish, or even disappear.

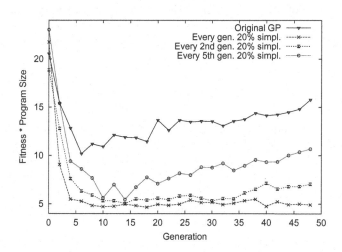

Fig. 4. Performance of the best fitness producing program

When choosing the probability of simplification, one has to make a trade-off between accuracy and program size:

- more accurate programs that grow moderately (less simplifications); or
- less accurate programs that do not grow (more simplifications).

5   Application

Our interest for limiting code growth in symbolic regression problems stems from the need for reducing the CPU time of our machine learning system [3]. The system is based on constructive induction, having the two components:

- learning engine - the C4.5 decision tree learning program [10]; and
- new feature generator - genetic programming.

We use this system in solving a mechanical engineering design problem, namely four bar mechanism synthesis [11]. The learning task is to discover the structural description of classes of such mechanisms. When generating the structural description, genetic programming creates the new features as algebraic functions of the six structural parameters of four bar mechanisms (simple symbolic regression with a terminal set consisting of six variables).

Genetic programming is called at the creation of each node in the decision tree (about 50 times in each run), on the data that has to be classified at that node . The training data set (i.e., the fitness cases) contains more than 7000 items. Thus, evaluating irrelevant code on these data takes much CPU time.

Initially, we tried to solve the problem by just using genetic programming, but the results were poor (there were too many errors in the produced description). The results produced by decision tree learning were better, but in many cases the original attributes could not describe correctly the classes. By allowing the creation of new attributes (algebraic expressions over the original attributes) at each node of the decision tree, the performance of the system improved considerably (from 47 rules and 6 misclassified cases when using C4.5 alone to 37 rules and 4 misclassified cases when using C4.5 with genetic programming at node level).

6   Conclusions

In the paper a method for limiting code growth in genetic programming was presented. The mechanism for controlling the size of programs was distinct from the fitness-based selection mechanism. The control of code growth was realized by a special mutation operator, inspired by three forms of mutation in biology. The method was applied to symbolic regression, where the special mutation operator consisted in algebraic simplification. The evolution of programs had two alternating phases:

- classical genetic programming - allowing code growth and intron formation (several generations); and
- simplification of programs - eliminating introns and reducing program size by means of a special mutation operator (one generation).

The experimental results show that code growth can be moderated or even stopped without the deterioration of performance by choosing the right frequency and probability for the application of simplification.

Acknowledgements. This work has been supported by grants No. T023305 and T25471 of the National Research Foundation of Hungary. The author is grateful to A. Márkus and J. Váncza for many helpful discussions.

References

1. Peter J. Angeline. Genetic programming and emergent intelligence. In Kenneth E. Kinnear, editor, *Advances in Genetic Programming*, pages 75–97. MIT Press, 1994.

2. Wolfgang Banzhaf, Peter Nordin, Robert E. Keller, and Frank D. Francone. *Genetic Programming: An Introduction.* Morgan Kaufmann, 1998.
3. Anikó Ekárt. Generating class descriptions of four bar linkages. In John R. Koza, editor, *Late Breaking Papers at the Genetic Programming 1998 Conference,* pages 42–47, 1998.
4. Walter Gilbert. Genes-in-pieces revisited. *Science,* 228:823–824, 1985.
5. Dale C. Hooper and Nicholas S. Flann. Improving the accuracy and robustness of genetic programming through expression simplification. In John R. Koza, David E. Goldberg, David B. Fogel, and Rick L. Riolo, editors, *Genetic Programming 1996: Proceedings of the First Annual Conference,* page 428, 1996.
6. Hitoshi Iba, Hugo de Garis, and Taisuke Sato. Genetic programming using a minimum description length principle. In Kenneth E. Kinnear, editor, *Advances in Genetic Programming,* pages 265–284. MIT Press, 1994.
7. John R. Koza. *Genetic Programming: On the Programming of Computers by Means of Natural Selection.* MIT Press, 1992.
8. William B. Langdon. Size fair and homologous tree genetic programming crossovers. In W. Banzhaf, J. Daida, A. E. Eiben, M. H. Garzon, V. Honavar, M. Jakiela, and R. E. Smith, editors, *GECCO-99: Proceedings of the Genetic and Evolutionary Computation Conference,* 1999.
9. Peter Nordin, Frank D. Francone, and Wolfgang Banzhaf. Explicitly defined introns and destructive crossover in genetic programming. In Justinian P. Rosca, editor, *Proceedings of the Workshop on Genetic Programming: From Theory to Real-World Applications,* pages 6–22, 1995.
10. J. Ross Quinlan. *C4.5: Programs for Machine Learning.* Morgan Kaufmann, 1993.
11. George N. Sandor and Arthur G. Erdman. *Advanced mechanism design: Analysis and synthesis,* volume 2. Prentice Hall, 1984.
12. Terence Soule, James A. Foster, and John Dickinson. Code growth in genetic programming. In John R. Koza, David E. Goldberg, David B. Fogel, and Rick L. Riolo, editors, *Genetic Programming 1996: Proceedings of the First Annual Conference,* pages 215–223, 1996.
13. Lubert Stryer. *Biochemistry.* Freeman, 1995.
14. J. Watson, N. H. Hopkins, J. W. Roberts, J. Argetsinger-Steitz, and A. M. Weiner. *Molecular Biology of the Gene.* Benjamin-Cummings, 1987.
15. Allan C. Wilson. The molecular basis of evolution. *Scientific American,* 253(4):148–157, 1985.
16. Byoung-Tak Zhang and Heinz Mühlenbein. Balancing accuracy and parsimony in genetic programming. *Evolutionary Computation,* 3(1):17–38, 1995.

Modeling and Analysis of Genetic Algorithm with Tournament Selection [*]

Anton V. Eremeev

Omsk Branch of Sobolev Institute of Mathematics,
13 Pevtsov st. 644099, Omsk, Russia.

Abstract. In this paper we propose a mathematical model of a simplified version of genetic algorithm (GA) based on mutation and tournament selection and obtain upper and lower bounds on expected proportion of the individuals with the fitness above certain threshold. As an illustration we consider a GA optimizing the bit-counting function and a GA for the vertex cover problem on graphs of a special structure. The theoretical estimates obtained are compared with experimental results.

1 Introduction

In this paper we propose a mathematical model of a simplified version of the genetic algorithm (GA) based on mutation and selection operators and evaluate the probability distribution of its population. We study the GA with s-tournament selection operator which randomly chooses s individuals from the previous population and selects the best one of them (see e.g. [4,10]).

The predictions of GA behavior are based on some a priori known parameters of the mutation operator. Using the proposed model we obtain upper and lower bounds on expected proportion of the individuals with fitness above certain thresholds. These bounds resemble the well-known Schema Theorem (see e.g. [3]), but the notion of schema is not used in this model. Instead of schemata here we consider the sets of genotypes with the fitness bounded from below.

In this framework we analyze the GA optimizing the bit-counting function and a GA for the vertex cover problem on graphs of a special structure. Finally the theoretical predictions are compared with the results of computational experiments.

Let the optimization problem consist in maximization of the goal function $f : X \to \mathbf{R}$, where X is the space of solutions. The GA searches for the optimal and sub-optimal solutions using a population of individuals, where each individual consists of genotype g and phenotype $x(g) \in X$. Here g is a fixed length string of genes $g^1, g^2, \ldots, g^n$, and the genes $g^i, i = 1, 2, ..., n$ are the symbols of a certain alphabet A (the alphabet $A = \{0, 1\}$ is the most commonly used). The function $x(g)$ maps genotype g to X, thus defining the representation of solutions in GA.

[*] The research was supported in part by the Russian Fund for Basic Research Grant 97-01-00771.

In each iteration the GA constructs a new population on the basis of the previous one. The search process is guided by the values of nonnegative fitness function $\Phi(g) = \phi(f(x(g)))$, which defines the fitness of the individual with genotype g. Here $\phi : R \to R$ is a monotone function, which is usually used for tuning the GA for solving a particular class of problems.

The individuals of the population may be ordered according to the sequence in which they are generated, thus the population may be considered as a vector of genotypes $\Pi^t = (g_1^{(t)}, g_2^{(t)}, ..., g_N^{(t)})$, where N is the size of population, which is constant during the run of the GA, and t is the number of the current iteration. In this paper we consider the populational version of GA, where all individuals of a new population are generated independently from each other with identical probability distribution depending on the existing population only.

Each individual is generated through selection of a parent genotype by means of selection operator, and modification of this genotype in mutation operator. During the mutation a subset of genes in the genotype string g is randomly altered. In general the mutation operator may be viewed as a random variable $Mut(g) \in A^n$ (where A^n is the set of all genotype strings) with the probability distribution depending on g. The most frequently used type of this operator randomly changes each gene of g with a fixed mutation probability p_m.

The genotypes of the initial population Π^0 are generated with some a priori chosen probability distribution. The stopping criterion is usually the limit of maximum iterations t_{max}. The result is the best solution generated during the run. In this paper we will consider a GA with the following scheme.

1. Generate the initial population Π^0.
2. For $t := 0$ to $t_{max} - 1$ do
    2.1. For $k := 1$ to N do
        Choose the parent genotype g from Π^t using s-tournament selection.
        Mutate g.
        Add $g_k^{(t+1)} = Mut(g)$ to the population Π^{t+1}.

The s-tournament selection operator randomly chooses s individuals from the previous population and selects the best one of them. In literature there are other standard operators, besides the tournament selection, in use. For example, the proportional selection where the selection probability is proportional to the fitness of individuals [3], or the truncation selection, where the parents are randomly chosen from $T\%$ best individuals of the previous population [7].

The GA may be considered as a Markov chain in a number of ways. For example, the states of the chain may correspond to different vectors of N genotypes that constitute the population Π^t. In this case if the genotype consists of n genes from the alphabet $A = \{0, 1\}$ then the number of states in the Markov chain is 2^{nN}. With the help of this model for the GA with proportional selection the following result is obtained [9].

Theorem 1. *Let x^t be the best solution found by GA until the step t; $f^* = \max\{f(x(g)) : g \in \{0,1\}^n\}$, and the probability of changing a gene in mutation is $p_m \in (0,1)$. Then $\lim_{t\to\infty} P\{f(x^t) = f^*\} = 1$.*

It is also proven that if this GA worked for infinite number of iterations, the genotypes corresponding to the optimal solutions would be lost and found infinitely often (see [9]). The similar results may be obtained for the GAs with tournament selection.

Another model representing the GA as a Markov chain has been proposed in [8], where all populations which differ only in the ordering of individuals are considered to be equivalent. In case if $A = \{0, 1\}$, each state of this Markov chain may be represented by a vector of 2^n components, where the proportion of each genotype in the population is indicated by the corresponding coordinate. In the framework of this model, M.Vose and collaborates have obtained a number of general results concerning the emergent behavior of GA [8,11,12].

The major difficulties in practical application of these models to real optimization problems are connected with the necessity to use the a priori information on fitness value of each genotype. Besides that, the size of the transition matrix grows exponentially as the length of genotype increases. In this paper we will consider one of the possible ways to handle these difficulties. The way we will use is the grouping of the states of population into larger classes.

2   The Approximating Model

The model proposed here represents the GA with tournament selection as it was introduced earlier. Here the information about the fitness of separate individuals is not used explicitly. Instead, the model makes use of certain a priori known parameters of probability distribution of the mutation operator described below.

Assume that there are d level lines of the fitness function fixed such that $\Phi_0 = 0 < \Phi_1 < \Phi_2 \ldots < \Phi_d$. The number of level lines and the fitness values corresponding to them may be chosen arbitrarily, but they should be relevant to the given problem and the mutation operator to yield a meaningful model. Let us introduce the following sequence of subsets of the set A^n:

$$H_i = \{g : \Phi(g) \geq \Phi_i\}, \quad i = 0, \ldots, d.$$

Due to nonnegativity of the fitness function, H_0 equals to the set of all genotypes. Besides that, for the sake of convenience, let us define the set $H_{d+1} = \emptyset$.

Now suppose that for all $i = 0, ..., d$ and $j = 1, ..., d$ the a priori lower bounds α_{ij} and upper bounds β_{ij} on mutation transition probability from subset $H_i \setminus H_{i+1}$ to H_j are known, i.e. for every $g \in H_i \setminus H_{i+1}$ holds $\alpha_{ij} \leq P\{Mut(g) \in H_j | g\} \leq \beta_{ij}$, where $P\{Mut(g) \in H_j | g\} = \sum_{g' \in H_j} P\{Mut(g) = g' | g\}$.

Let $\mathbf{A}$ denote the matrix with the elements α_{ij} where $i = 0, ..., d$, and $j = 1, ..., d$. The similar matrix of upper bounds β_{ij} is denoted by $\mathbf{B}$. Let the population on iteration t be represented by the *population vector* $z^{(t)} = (z_1^{(t)}, z_2^{(t)}, \ldots, z_d^{(t)})$ where $z_i^{(t)} \in \mathbf{R}$ is the proportion of genotypes from H_i within the population Π^t. The vector $z^{(t)}$ is a random vector, where $z_i^{(t)} \geq z_{i+1}^{(t)}$ for $i = 1, ..., d-1$ since $H_{i+1} \subseteq H_i$. Let $P\{g^{(t)} \in H_j\}$ be the probability that an individual, which is added after selection and mutation into Π^t, has a geno-type from H_j for $j = 0, ..., d$, and $t > 0$ (according to the scheme of the algo-rithm this probability is identical for all genotypes of Π^t, i.e. $P\{g^{(t)} \in H_j\} = P\{g_1^{(t)} \in H_j\} = ... = P\{g_N^{(t)} \in H_j\}$). Let E denote the mathematical expecta-tion. Then it is not difficult to obtain the following:

Proposition 1. $E[z_i^{(t)}] = P\{g^{(t)} \in H_i\}$ *for all* $t > 0, i = 1, ..., d$.

3   Bounding the Expectation of Population Vector

In this section the lower and upper bounds on proportion of genotypes from H_i for all $i = 1, ..., d$ will be considered. Let $P_{ch}(S, z)$ denote the probability that the genotype, chosen by the tournament selection from the current population with vector z, belongs to a subset $S \subseteq A^n$. Note that if the current population is represented by the vector $z^{(t)} = z$, then a genotype obtained by selection and mutation would belong to H_j with conditional probability

$$P\{g^{(t+1)} \in H_j | z^{(t)} = z\} = \sum_{i=0}^{d} \sum_{g \in H_i \setminus H_{i+1}} P\{Mut(g) \in H_j | g\} P_{ch}(\{g\}, z) \geq$$

$$\sum_{i=0}^{d} \alpha_{ij} \sum_{g \in H_i \setminus H_{i+1}} P_{ch}(\{g\}, z) = \sum_{i=0}^{d} \alpha_{ij} P_{ch}(H_i \setminus H_{i+1}, z). \tag{1}$$

Given the tournament size s we obtain the following selection probabilities: $P_{ch}(H_i, z^{(t)}) = 1 - (1 - z_i^{(t)})^s$, and, consequently, $P_{ch}(H_i \setminus H_{i+1}, z) = P_{ch}(H_i, z) - P_{ch}(H_{i+1}, z) = (1 - z_{i+1})^s - (1 - z_i)^s$. This leads to the inequality:

$$P\{g^{(t+1)} \in H_j | z^{(t)} = z\} \geq \sum_{i=0}^{d} \alpha_{ij}((1 - z_{i+1})^s - (1 - z_i)^s).$$

Let $Z_N = \{z \in \mathbf{R}^d : z_i \in \{0, \frac{1}{N}, \frac{2}{N}, ..., 1\}, z_i \geq z_{i+1}\}$ be the set of all possi-ble vectors of population which consists of N individuals. Then using the total probability formula we obtain the following bound on unconditional probability:

$$P\{g^{(t+1)} \in H_j\} \geq \sum_{z \in Z_N} \sum_{i=0}^{d} \alpha_{ij}((1 - z_{i+1})^s - (1 - z_i)^s)P\{z^{(t)} = z\} =$$

$$\sum_{i=0}^{d} \alpha_{ij} E[(1 - z_{i+1}^{(t)})^s - (1 - z_i^{(t)})^s]. \tag{2}$$

The Proposition 1 yields that $E[z_j^{(t+1)}] = P\{g^{(t+1)} \in H_j\}$. Consequently since $(1 - z_{d+1}^{(t)})^s = 1$ and $(1 - z_0^{(t)})^s = 0$,

$$E[z_j^{(t+1)}] \geq \alpha_{dj} - \sum_{i=1}^{d}(\alpha_{i,j} - \alpha_{i-1,j})E[(1 - z_i^{(t)})^s]. \tag{3}$$

Denote $I_j^+(\mathbf{A}) = \{1 \leq i \leq d : \alpha_{i,j} - \alpha_{i-1,j} \geq 0\}$ and $I_j^-(\mathbf{A}) = \{1 \leq i \leq d : \alpha_{i,j} - \alpha_{i-1,j} < 0\}$. Let us apply the Jensen inequality to those terms of sum (3) for which $i \in I_j^-(\mathbf{A})$ (in view of the fact that under the expectation sign the function on $z_i^{(t)}$ is convex). The terms where $i \in I_j^+(\mathbf{A})$ can be bounded using the simple estimate $(1 - z_i^{(t)})^s \leq 1 - z_i^{(t)}$. This leads to the following proposition.

Proposition 2. *The components $E[z_j^{(t+1)}]$ of expected next population vector are bounded below for all $j = 1, ..., d$ and $t \geq 0$ as follows*

$$E[z_j^{(t+1)}] \geq \alpha_{dj} - \sum_{I_j^+(\mathbf{A})} (\alpha_{ij} - \alpha_{i-1,j})(1 - E[z_i^{(t)}]) - \sum_{I_j^-(\mathbf{A})} (\alpha_{ij} - \alpha_{i-1,j})(1 - E[z_i^{(t)}])^s.$$

$$(4)$$

Note that if the probability $P\{Mut(g) \in H_j|g\}$ for all $i = 0, ..., d$, and $j = 1, ..., d$ does not depend on choice of $g \in H_i \backslash H_{i+1}$, then we may assume that $\alpha_{ij} = \beta_{ij} = P\{Mut(g) \in H_j|g \in H_i \backslash H_{i+1}\}$ for all i and j. In this case the mutation operator will be called a *step mutation operator with respect to the sequence of subsets $H_0, H_1, ..., H_d$* (or a step mutation operator for short).

In order to emphasize the fact that $\gamma_{ij} = P\{Mut(g) \in H_j|g \in H_i \backslash H_{i+1}\}$ is not the transition probability used in the Markov chains (which is not considered in this paper) sometimes we will call γ_{ij} the *threshold transition probability*. The matrix $\mathbf{\Gamma}$ will detone here the matrix of threshold transition probabilities of a step mutation operator: $\mathbf{\Gamma} = \mathbf{A} = \mathbf{B}$.

If the tournament size $s = 1$, then the selection has a uniform distribution and its operation does not depend on the fitness of individuals. It is easy to see that in this special case (4) becomes tight for the GA with step mutation operator. Our aim in this section is to obtain a lower bound on $E[z^{(t)}]$ for arbitrary t if the expectation of the initial population vector $E[z^{(0)}]$ is given. In order to do this let us introduce the following definition.

A matrix $\mathbf{M}$ with elements $m_{ij}, i = 0, ..., d$, and $j = 1, ..., d$ will be called *monotone* if $m_{i-1,j} \leq m_{i,j}$ for all i, j from 1 to d.

The matrix of bounds on transition probabilities is monotone if for any $j = 1, ..., d$ the genotypes from any subset H_i have the bound on transition probability to H_j not less than the bounds of the genotypes from the subsets $H_{i'}$ for all $i' < i$. Obviously, for any mutation operator the monotone bounds exist. (For example $\mathbf{A} = \mathbf{0}$ where $\mathbf{0}$ is a zero matrix and $\mathbf{B} = \mathbf{U}$ where $\mathbf{U}$ is the matrix with all elements equal 1). The problem may be connected only with the absence of bounds which are sharp enough to evaluate the mutation operator properly. Imposing the assumption of monotone bounds on (4) we derive the following :

Proposition 3. *If the matrix $\mathbf{A}$ is monotone, then for any tournament size $s \geq 1$ and $j = 1, \ldots, d$*

$$E[z_j^{(t+1)}] \geq \alpha_{0j} + \sum_{i=1}^{d} (\alpha_{i,j} - \alpha_{i-1,j}) E[z_i^{(t)}].$$

$$(5)$$

Let $\mathbf{W}$ be a $(d \times d)$ matrix with the elements $w_{ij} = \alpha_{ij} - \alpha_{i-1,j}$; $\mathbf{I}$ is the identity matrix of the same size, and vector $\alpha = (\alpha_{01}, ..., \alpha_{0d})$.

Theorem 2. *If the matrix $\mathbf{A}$ is monotone and $\alpha_{dj} - \alpha_{0j} < 1$ for $j = 1, ..., d$, then for all $t \geq 0$*

$$E[z^{(t)}] \geq E[z^{(0)}]\mathbf{W}^t + \alpha(\mathbf{I} - \mathbf{W})^{-1}(\mathbf{I} - \mathbf{W}^t). \tag{6}$$

Proof. Let us consider a sequence of d-dimensional vectors $y^{(0)}, y^{(1)}, ..., y^{(t)}, ...,$ where $y^{(0)} = E[z^{(0)}]$, $y^{(t+1)} = y^{(t)}\mathbf{W} + \alpha$. Note that the right-hand side of (5) will not increase if the components of $E[z^{(t)}]$ are substituted with their lower bounds. Thus by induction on t we get: $E[z^{(t)}] \geq y^{(t)}$ for any t.

Consider the vector norm $||z|| = \max_j |z_j|$ in $\mathbf{R}^d$ and the matrix norm $||W|| = \max_j \sum_{i=1}^{d} |w_{ij}|$ corresponding to it. Under the conditions of this theorem we have $w_{ij} = \alpha_{ij} - \alpha_{i-1,j} \geq 0$, and $||W|| = \max_j \sum_{i=1}^{d} w_{ij} = \max_j(\alpha_{dj} - \alpha_{0j}) < 1$. Therefore by the properties of linear operators we conclude that the matrix $(\mathbf{I}-\mathbf{W})^{-1}$ exists. Using the induction on t we obtain the identity: $y^{(t)} = y^{(0)}\mathbf{W}^t + \alpha(\mathbf{I} - \mathbf{W})^{-1}(\mathbf{I} - \mathbf{W}^t)$, which leads to (6). $\square$

Note that in most of the GA implementations an arbitrary given genotype may be produced with a non-zero probability as a result of mutation, and the corresponding Markov chain is ergodic (see e.g. [9]). In this case the condition $\alpha_{dj} - \alpha_{0j} < 1$ is obviously satisfied for all j, if the bounds are properly chosen.

As it follows from the proof of the Theorem 2, $||W^t|| \leq ||W||^t < 1$ and (6) approaches $\alpha(\mathbf{I} - \mathbf{W})^{-1}$ when t tends to infinity, thus the limit of this bound does not depend on distribution of the initial population. Also let us note that the inequality (6) turns into equation in the case of a step mutation operator and the tournament size $s = 1$.

By reasoning similar to the proof of Proposition 2 we obtain the following:

Proposition 4. *The components of expected next population vector are bounded above as follows*

$$E[z_j^{(t+1)}] \leq \beta_{dj} - \sum_{I_j^-(\mathbf{B})} (\beta_{ij} - \beta_{i-1,j})(1 - E[z_i^{(t)}]) - \sum_{I_j^+(\mathbf{B})} (\beta_{ij} - \beta_{i-1,j})(1 - E[z_i^{(t)}])^s$$

$$\tag{7}$$

for $j = 1, ..., d$, and if the matrix $\mathbf{B}$ is monotone then

$$E[z_j^{(t+1)}] \leq \beta_{dj} - \sum_{i=1}^{d}(\beta_{ij} - \beta_{i-1,j})(1 - E[z_i^{(t)}])^s. \tag{8}$$

By means of iterative application of (8) the components of the vectors $E[z^{(t)}]$ may be bounded up to arbitrary t, starting from the expectation of the initial population vector $E[z^{(0)}]$. The nonlinearity in the right-hand side of (8), however, creates an obstacle for obtaining an analytical bound similar to the bound (6) of Theorem 2. This problem could be tackled using the contractive transformations theory or by approximation of the corresponding difference equation with the help of differential one, but this investigation is outside the scope of this paper.

Note that all of the bounds obtained up to this point do not include the population size and they are valid for arbitrary N. The Theorem 3 in the following section will show that the right-hand side of (8) reflects the asymptotic behavior of population under step mutation operator as $N \to \infty$.

3.1 GA with a Step Mutation Operator

If the GA uses a step mutation operator, the probability distribution of the next population is completely determined by the vector of the current population. In this case the GA may be viewed as a Markov chain with the states corresponding to the elements of Z_N.

Note that in general the population vectors are random values depending on N. In order to express this fact in notation let us denote the proportion of genotypes from H_i in population Π^t by $z_i^t(N)$.

Theorem 3. *Let the GA use a step mutation operator with monotone threshold transition matrix* $\mathbf{\Gamma}$, *and let the genotypes of the initial population be identically distributed. Assume that the sequence of d-dimensional vectors* $y^{(0)}, y^{(1)}, ..., y^{(t)}, ...$ *is defined as follows:*

$$y^{(0)} = E[z^{(0)}(N)], \qquad y_j^{(t+1)} = \gamma_{dj} - \sum_{i=1}^{d}(\gamma_{ij} - \gamma_{i-1,j})(1 - y_i^{(t)})^s \qquad (9)$$

for $j = 1, ..., d$ and $t \geq 0$. Then $E[z^{(t)}(N)] \xrightarrow[N \to \infty]{} y^{(t)}$ *at any iteration t.*

Proof. Since the GA is based on a step mutation operator,

$$E[z_j^{(t+1)}(N)] = \gamma_{dj} - \sum_{i=1}^{d}(\gamma_{ij} - \gamma_{i-1,j})E[(1 - z_i^{(t)}(N))^s].$$

Consequently if it is proven that

$$\lim_{N \to \infty}\left(E[(1 - z_i^{(t)}(N))^s] - (1 - E[z_i^{(t)}(N)])^s\right) = 0, \qquad (10)$$

then the convergence of $E[z^{(t)}(N)]$ to $y^{(t)}$ will mean that $E[z^{(t+1)}(N)]$ converges to $y^{(t+1)}$ as $N \to \infty$. In this case the statement of the theorem follows by induction on t for arbitrary finite t.

In order to prove (10) let us fix $t \geq 0$ and consider the sequence of independent identically distributed random variables $\xi_1^i, \xi_2^i, ..., \xi_N^i$, where $\xi_l^i = 1$ if the genotype of l-th individual in population $\Pi^{(t)}$ belongs to H_i, and $\xi_l^i = 0$ otherwise.

Using the law of large numbers, for any $i = 1, ..., d$ and $\varepsilon > 0$ we obtain

$$P\left\{\left|\frac{\sum_{l=1}^{N}\xi_l^i}{N} - E[\xi_1^i]\right| < \varepsilon\right\} \xrightarrow[N \to \infty]{} 1.$$

Note that $\sum_{l=1}^{N} \xi_l^i / N = z_i^{(t)}(N)$, and besides that, in view of Proposition 1, $E[\xi_1^i] = P\{\xi_1^i = 1\} = E[z_i^{(t)}(N)]$ (in case if $t = 0$ this equality also holds, since the genotypes of the initial population are identically distributed). Consequently, $P\left\{\left|z_i^{(t)}(N) - E[z_i^{(t)}(N)]\right| < \varepsilon\right\} \xrightarrow[N\to\infty]{} 1$ for any $\varepsilon > 0$. Hence, by continuity of the function $(1-x)^s$, it follows that

$$P\left\{\left|(1 - z_i^{(t)}(N))^s - (1 - E[z_i^{(t)}(N)])^s\right| \geq \varepsilon\right\} \xrightarrow[N\to\infty]{} 0.$$

Let us denote $F_N(x) = P\left\{(1 - z_i^{(t)}(N))^s - (1 - E[z_i^{(t)}(N)])^s < x\right\}$. Then

$$\lim_{N\to\infty}\left(E\left[(1 - z_i^{(t)}(N))^s\right] - (1 - E[z_i^{(t)}(N)])^s\right) = \lim_{N\to\infty}\int_{-\infty}^{\infty} x\, dF_N(x) \leq$$

$$P\left\{\left|(1 - z_i^{(t)}(N))^s - (1 - E[z_i^{(t)}(N)])^s\right| \geq \varepsilon\right\} + \lim_{N\to\infty}\int_{|x|<\varepsilon} \varepsilon\, dF_N(x) \xrightarrow[N\to\infty]{} \varepsilon.$$

for arbitrary $\varepsilon > 0$, hence (10) holds. $\square$

Corollary 1. *Suppose that a step mutation operator with monotone threshold transition matrix is given. Let $z^{(t)}(N)$ be a population vector of GA with tournament size s, and let $\hat{z}^{(t)}(N)$ represent a population of GA with tournament size $\hat{s} \geq s$.*

If the individuals of initial populations are identically distributed, and $E[\hat{z}_i^{(0)}(N)] \geq E[z_i^{(0)}(N)]$ for $i = 1, ..., d$, then for any iteration t and $i = 1, ..., d$ $E[\hat{z}_i^{(t)}(N)] \geq E[z_i^{(t)}(N)]$ holds, provided that N is big enough.

4 Some Applications of the Model

First we shall consider the simple *bit-counting* fitness function $\Phi : \{0,1\}^n \to \{0, ..., n\}$, which equals the number of 1's in the binary string of genotype. For this problem a number of versions of GA with truncation selection have been considered in [1,5] and other papers. Suppose that the GA uses the standard mutation operator, changing every gene with probability p_m. Let the subsets $H_0, ..., H_d$ be defined by the level lines $\Phi_0 = 0, \Phi_1 = 1, ..., \Phi_d = d$ and $d = n$. It is easy to see that in this case the GA has a step mutation operator. The matrix Γ for this operator could be obtained using the result from [1], but here we shall consider this example as a special case of the following problem.

Let the representation of the problem admit a decomposition of the genotype string into d nonoverlapping substrings (called *blocks* here) in such a way that the fitness function Φ equals to the number of blocks for which a certain property **K** holds [1]. Let m be the number of blocks and let $K(g, \lambda) = 1$ if **K** holds for the block λ of genotype g, and $K(g, \lambda) = 0$ otherwise (here $\lambda = 1, ..., m$).

---

[1] These functions are a special case of the additively decomposed functions, where the elementary functions are boolean and substrings are nonoverlapping (see e.g. [6]).

Suppose that during mutation, any block for which $\mathbf{K}$ did not hold gets the property $\mathbf{K}$ with probability $\tilde{r}$, i.e. $P\{K(Mut(g), \lambda) = 1 | K(g, \lambda) = 0\} = \tilde{r}$ for $\lambda = 1, ..., m$. On the other hand, assume that a block with the property $\mathbf{K}$ keeps this property during mutation with probability r, i.e. $P\{K(Mut(g), \lambda) = 1 | K(g, \lambda) = 1\} = r; \lambda = 1, ..., m$. Let the subsets $H_0, ..., H_d$ correspond to the level lines $\Phi_0 = 0, \Phi_1 = 1, ..., \Phi_d = d$ again. In this case the element γ_{ij} of threshold transition matrix Γ equals the probability to obtain a genotype containing j or more blocks with property $\mathbf{K}$ after mutation of a genotype which contained i blocks with this property. It is not difficult to see that for this mutation operator the threshold transition probabilities $\gamma_{ij} = P\{Mut(g) \in H_j | g \in H_i \backslash H_{i+1}\}$ for $i = 0, ..., d, j = 1, ..., d$ are given by the following expression:

$$\gamma_{ij} = \sum_{k=0}^{d-i} \binom{d-i}{k} \tilde{r}^k (1-\tilde{r})^{d-i-k} \sum_{\nu=0}^{\min\{i, i-j+k\}} \binom{i}{\nu} (1-r)^\nu r^{i-\nu}. \tag{11}$$

Verifying the definition of monotone matrix we obtain the following:

Proposition 5. *If $r \geq \tilde{r}$ then the matrix Γ defined by (11) is monotone.*

Now for the bit-counting function the matrix Γ is obtained, assuming that $\tilde{r} = (1 - r) = p_m$, $d = n$. Obviously, this operator is monotone if $p_m \leq 0.5$.

The formula (11) may be also used for finding the threshold transition matrices of some other optimization problems with a "regular" structure. As an example we consider the vertex cover problem (VCP) on graphs of a special structure. In general, the vertex cover problem is formulated as follows.

Let $G = (V, E)$ be a graph with a set of vertices V and the edge set E. A subset $C \subseteq V$ is called a vertex cover of G if every edge has at least one endpoint in C. The vertex cover problem is to find a vertex cover C^* of minimal cardinality.

Let G_d be a graph consisting of d disconnected triangle subgraphs. Obviously, each triangle is covered optimally by two vertices and the redundant cover consists of three vertices. In spite of the simplicity of this problem, it is proven in [13] that the working time of some well-known algorithms of branch and bound type grows exponentially on d if they are applied to the VCP on graph G_d.

Suppose that the VCP is handled by the GA with *non-binary* representation (see e.g. [2]): each gene $g^i \in \{0, 1\}, i = 1, ..., |E|$ corresponds to an edge of G_d, assigning one of its endpoints which has to be included in the cover. The phenotype $C = x(g)$ is a cover, containing all vertices which are assigned by at least one of the genes. Let the mutation operator alter each gene with probability p_m. The natural way to choose the fitness function in this case is to assume $\Phi(g) = |V| - |x(g)|$. Then for G_d the fitness $\Phi(g)$ coincides with the number of optimally covered blocks in $C = x(g)$. Let the genes representing the same triangle constitute a single block, and let the property $\mathbf{K}$ imply that a block is optimally covered. Then by looking at all possible ways of producing the redundant covers of the triangle subgraph one can see that $\tilde{r} = 1 - p_m^3 - (1 - p_m)^3$, and $r = 1 - p_m(1 - p_m)^2 - p_m^2(1 - p_m)$. Using (11) we obtain the threshold transition matrix for this mutation operator. It is easy to verify that in this case the inequality $r \geq \tilde{r}$ holds for any mutation probability p_m, and therefore the operator is always monotone.

4.1   Computation Experiments

This section presents some experimental results in comparison with the theoretical estimates obtained above. Here we will consider the application of GA to the VCP on graph G_d as it was described before. The proportion of the optimal genotypes in the population for GAs with different population size is presented in Fig.1. Here $d = 8$, $p_m = 0.01$, $s = 2$ and $z^{(0)} = 0$ (i.e. the initial population consists of genotypes that define a cover where each triangle subgraph is covered redundantly).

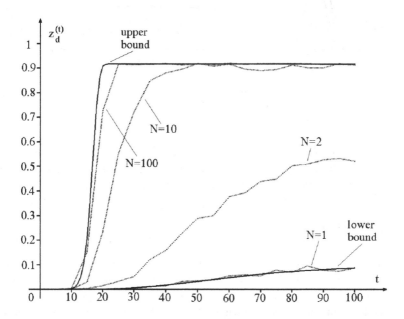

Fig. 1. Average proportion of optimal solutions as a function of the iteration number. $d = 8$, $p_m = 0.01$, $s = 2$.

The computational results are shown in dotted lines. The solid lines correspond to the lower and upper bounds given by (6) and (8). This plot shows that the upper bound (8) provides a good approximation to the value of $z_d^{(t)}$ obtained experimentally, even if the population size is not very large. The rest of the components of $z^{(t)}$ demonstrated a similar tendency.

Another series of experiments was carried out in order to compare the behavior of GAs with different tournament sizes. Figure 2 presents the results for the GA with $p_m = 0.1$, $N = 100$ and $z^{(0)} = 0$ solving the VCP on graph G_6.

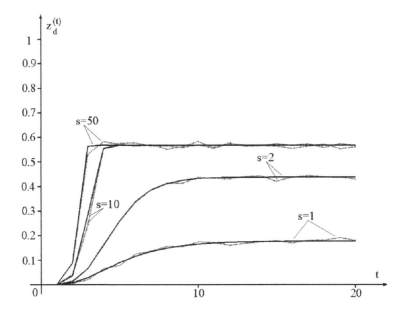

Fig. 2. Average proportion of optimal solutions as a function of the iteration number. $d = 6$, $p_m = 0.1$, $N = 100$.

This plot demonstrates the increase in proportion of the optimal genotypes caused by the extension of the tournament size, which is consistent with Theorem 3 and Corollary 1.

5   Conclusions

In this paper we presented a new model of GA with tournament selection and obtained the upper and lower bounds on proportion of "good" genotypes in population. The tournament selection is frequently used in applications, however theoretically it has not been investigated as much as the proportional selection.

The model may be applied to the GAs with different fitness functions and mutation operators. The adequacy of the bounds depends on the tightness of the monotone bounds on mutation transition probabilities. The analysis of interconnections between the mutation operator, the problem coding, and the goal function is separated from the rest of the GA model. This analysis of mutation ought to provide some coarse graining of the fitness landscape, and the usefulness of the model will depend on the precision of this information.

The bounds obtained provide the estimates for the speed of spreading of "good" genotypes in the population of GA. In particular, it is proved for the GA, which uses a step mutation operator with monotone threshold transition matrix, that the expected components of population vector reach the lower bounds if the population size $N = 1$, and they tend to the upper bounds if $N \to \infty$.

Further research is expected to involve the investigation of the crossover operator and applications of the model to more complex optimization problems.

Acknowledgment. The author would like to thank Prof. Boris A. Rogozin and Sergey A. Klokov for the helpful comments on this paper.

References

1. Bäck, T.:The interaction of mutation rate, selection, and self-adaptation within a genetic algorithm. In: Proc. of Parallel Problem Solving from Nature (PPSN II) Männer, R. and Manderick, B. (eds.). North Holland (1993) 85–94
2. Eremeev, A.V.: A genetic algorithm with a non-binary representation for the set covering problem. In: Proc. of Operations Research (OR'98). Springer-Verlag (1999) 175–181
3. Goldberg, D.E.: Genetic algorithms in search, optimization and machine learning. Addison Wesley Publishing Company (1989)
4. Goldberg, D.E., Korb, B. and Deb, K.: Messy genetic algorithms: motivation, analysis, and first results. Complex Systems, **3** (5) (1989) 493–530
5. Mühlenbein, H.: How genetic algorithms really work: I. Mutation and hillclimbing. In: Proc. of Parallel Problem Solving from Nature (PPSN II) Männer, R. and Manderick, B. (eds.). North Holland (1993) 15–26
6. Mühlenbein, H., Mahnig, T. and Rodriguez, A. O.: Schemata, distributions and graphical models in evolutionary optimization. To Appear in Journal of Heuristics.
7. Mühlenbein, H. and Schlierkamp-Voosen, D.: The science of breeding and its application to the breeder genetic algorithm (BGA). Evolutionary Computation, **1** (1) (1993) 335–360
8. Nix, A. and Vose, M. D.: Modeling genetic algorithms with Markov chains. Annals of Mathematics and Artificial Intelligence, **5** (1992) 79–88
9. Rudolph, G.: Convergence analysis of canonical genetic algorithms. IEEE Transactions on Neural Networks, **5** (1) (1994) 96–101
10. Thierens, D. and Goldberg, D.E.: Convergence models of genetic algorithm selection schemes In: Proc. of Parallel Problem Solving from Nature (PPSN III) Davidor, Y., Schwefel, H.- P. and Männer, R. (eds.). Springer-Verlag (1994) 117–129
11. Vose, M. D.: Modeling simple genetic algorithms. Evolutionary Computation, **3** (4) (1995) 453–472
12. Vose, M. D. and Wright, A.H.: The Walsh transform and the theory of the simple genetic algorithm. Genetic Algorithms for Pattern Recognition. Pal, S. K., Wang, P. P. (eds.) (1995) 25–44
13. Zaozerskaya, L.A.: Investigation and solving of some classes of integer programming problems on the basis of the regular partitions. PhD. thesis, Omsk Branch of Sobolev Institute of Mathematics, SB RAS. (1998)

On Generating HTML Style Sheets with an Interactive Genetic Algorithm Based on Gene Frequencies

Nicolas Monmarché[1], G. Nocent[1], Gilles Venturini[1], and P. Santini[2]

[1] Laboratoire d'Informatique, Université de Tours,
64, Avenue Jean Portalis, 37200 Tours, France.
monmarche, venturini@univ-tours.fr,
Phone: +33 2 47 36 14 14, Fax: +33 2 47 36 14 22

[2] Alderan,
221, avenue du Président Wilson
93218 La Plaine Saint Denis Cédex, France
info@Alderan.fr
Phone: +33 1 49 46 47 93, Fax: +33 1 49 46 47 99

Abstract. We present in this paper a new interactive method called Imagine that automatically generates style sheets for Web sites. This method aims at satisfying the artistic or aesthetic preferences of the user. This method uses a genetic algorithm to generate style sheets and to find in a search space one or several style sheets that will maximize the user satisfaction. This genetic algorithm is interactive: it generates style sheets, it displays them, and then it asks the user to select those which look the best. In this way, the search for an optimal sheet is guided by the answers provided by the user. Also, this algorithm uses non standard genetic operators based on gene frequencies. We present examples obtained with the actual prototype.

1   Introduction

Usually, one can formulate an optimization or learning problem by defining a space of solutions and a mathematical function, the so-called evaluation function, that must be maximized by finding the best solution in the space. As long as it is possible to evaluate automatically the quality of a solution using this evaluation function, then numerous optimization methods exist. However, there are some problems for which the quality of solutions cannot be formulated mathematically, because for instance this quality depends on artistic or aesthetic criteria that only the user can evaluate. Thus, there are some domains in which either there is no mathematical language to represent the user preferences, or the user is not definitely sure about what he really wants and would like to let the computer make some propositions.

The problem that we deal with and solve in this paper has exactly this property. When one wants to define a "nice" style sheet for HTML pages and

C. Fonlupt et al. (Eds.): AE'99, LNCS 1829, pp. 99–110, 2000.

for a given WWW site, it is impossible to mathematically quantify what a "nice style sheet" means. This depends highly on the user, on the content of the site, and on the possibilities offered by the HTML. Furthermore, this aesthetic aspect is obviously of high importance in a Web site because it contributes to its look and finally to how the site will be perceived by visitors or customers. Actual commercial tools for editing HTML pages (such as FrontPage, Netscape Composer, VisualPage) require that the user specifies by himself many elements of crucial importance for the final look of his or her site.

To improve these HTML page conception tools, we present a new method which is much more interactive. This method can autonomously propose solutions that the user may not have thought about, and can take into account the user's preferences. It is based on an interactive genetic algorithm (IGA) and uses the following principles: an initial population of style sheets is generated randomly. Then these sheets are presented to the user in a graphical way, by applying the style they represent to one or more HTML pages that were initially provided by the user (and which represent the Web site to optimize). The user can thus select the style sheets that he favors. The characteristics of the selected sheets are taken into account by the genetic algorithm (GA) in order to compute the next generation of style sheets. These new individuals will be presented again to the user, and this cycle is repeated until the selection, recombination and mutation of style sheet characteristics produce a sheet that satisfies the user. The interactive selection of style sheets based on visualization is the mean by which the user's preferences are highlighted and taken into account in a simple and intuitive way. The genetic recombination and mutation is the metaheuristic used to take into account the user's preferences and to propose a new generation without any other help from the user.

The remaining of this paper is organized as follows: section 2 describes the main principles of the IGAs. Section 3 presents the genetic representation of the problem: how to represent style sheets, how to define genetic operators, how to select individuals. Section 4 presents the actual prototype and examples of results obtained with real pages. Section 5 concludes on present and future work.

2   Overview of Interactive Genetic Algorithms

Recently, it has been shown how the evaluation function in genetic algorithms (Holland 1975) (Michalewicz 1992) can be replaced by a user-based evaluation of individuals. These so-called interactive genetic algorithms allow the user to directly select the individuals that he or she favors (Dawkins 1986) (Todd et Latham 1991) (Caldwell et Johnston 1991) (Sims 1991) (Smith 1991) (Venturini et Montalibet 1995) (Graf et Banzhaf 1996) (Venturini et al. 1997). IGAs lead to new kind of applications in computer science like the reconstruction of a criminal face guided by the victim, or the creation of beautiful images. In those applications, the user selects individuals according to his or her own criteria and the IGA evolves the individuals according to the user's preferences.

The two main conditions required to use an IGA in problem solving are the following ones: the evaluation function must be difficult to define in an automatic and mathematical way. This is the case of most domains where the user must be involved in the obtained results and must formulate his or her own preferences. The second condition is that the user must be able to easily evaluate the individuals, which generally implies to visualize the individuals.

These two conditions are fulfilled in our style sheet generation problem and are especially interesting for this problem. The user must be involved in the optimization process in order to let him or her formulate preferences for a given style sheet, and it is possible to visualize those style sheets using for instance an HTML browser.

3   Imagine : An IGA for HTML Style Sheet Generation

3.1   Genetic Representation of the Problem

Table 1. The 5 genes related to the global characteristics of an HTML page. The list of values are non exhaustive because the user may add more values or files (but also remove some of them).

Genes related to an HTML document	Possible values
Background (image or color)	64 .gif or .jpeg textures and 128 colors
Color of links	128 RGB triples
Rules	25 .gif or .jpeg files + 1 predef. std value
Bullets	25 .gif or .jpeg files + 4 predef. std values
Arrows ("Back", "Next", "Home")	25 triples of .gif or .jpeg files

We consider in this paper that the user would like to create HTML pages that partially or entirely represent a web site. We assume that the content of the site has been determined but that its appearance must be optimized in interaction with the user. Among all the elements that contribute to the look of a web site, we have selected the elements mentioned in tables 1 and 3.1. In table 1, we have represented the main general elements which can be used to determine a general style of the pages, and more precisely, the background, the links, the bullets, the rules and the navigation arrows. In table 3.1, we have represented the elements that determine the appearance of the text and paragraphs. These 26 elements can be considered as genes and will form the genetic representation.

It is important to notice that, for several genes like textures, colors and more generally all Gif or Jpeg files that correspond to rules, bullets, etc, the user has the possibility to add or remove some values. In this way, our tool manages a kind of database which can be modified at will by the user before running the IGA. New elements can thus be added, and especially elements which are relevant in a given domain.

Table 2. The 21 genes related to the paragraph style.

Genes related to paragraphs	Values
Title (levels 1, 2 and 3): font color	128 RGB triples
Title (levels 1, 2 and 3): alignment	left, right, centered, justified
Title (levels 1, 2 and 3): background color	128 RGB triples
Title level 1: font	35 fonts
Title level 1: style	normal, italic, oblique
Title level 1: weight	lighter, normal, bold, bolder
Title level 2: font	35 fonts
Title level 2: style	normal, italic, oblique
Title level 2: weight	lighter, normal, bold, bolder
Title level 3: font	35 fonts
Title level 3: style	normal, italic, oblique
Title level 3: weight	lighter, normal, bold, bolder
Paragraph: font	35 fonts
Paragraph: style	normal, italic, oblique
Paragraph: weight	lighter, normal, bold, bolder
Paragraph: font color	128 RGB triples
Paragraph : alignment	left, right, centered, justified
Text of introduction: style	normal, italic, oblique
Text of introduction: weight	lighter, normal, bold, bolder
Comments: style	normal, italic, oblique
Comments: weight	lighter, normal, bold, bolder

Once these l ($= 26$ in the actual prototypes) elements have been defined, one can define the search space simply as a discrete space of l dimensions where each dimension corresponds to a gene i with k_i possible values. As mentioned previously, tables 1 and 3.1 are summing up these 26 genes and their possible values.

3.2   A Non Standard GA That Uses Gene Frequencies

We use in this paper a non standard GA for the following reason: when the user must evaluate the individuals, one must display the whole population of individuals. However, it is impossible to display more than 12 style sheets simultaneously without greatly decreasing the quality of the visualization, even on a 21" screen. One possibility to avoid this problem would be to display the population with several successive visualizations, but this is not very intuitive and does not allow the user to compare all individuals to each others. It is thus impossible to display a population of more than 12 individuals, while standard GAs usually require hundreds or thousands of individuals.

But recently, authors have shown that the standard model of GAs which evolves a finite population of individuals can be generalized to a model with similar or better performances but which makes use of an infinite population represented by a vector of probabilities (Syswerda 1993) (Baluja 1994) (Sebag and

Ducoulombier 1998). To explain briefly such principles, let us consider a gene i with k_i possible values and let $V_i = (V_i(1), \ldots, V_i(k_i))$ be the proportions of the gene i values in the current population. One can notice that one point crossover for instance does not modify these proportions, because it only exchanges gene values at the same location. So the crossover operator does only generate individuals according to these proportions. It may thus be simply replaced with a more general operator which could use $(V_i(1), \ldots, V_i(k_i))$ as a vector of probabilities, where $V_i(j)$ would denote the probability of generating the jth value of gene i. In the same way, the mutation operator can be viewed, for a given gene i, as a small perturbation of the probability vector V_i.

3.3 IGAs Principles

The GA used in this paper in described in figure 1. It explicitly manages a set of probability vectors which has been denoted by $(V_1, \ldots, V_l)$ of dimensions $(k_1, \ldots, k_l)$ and where $V_i = (V_i(1), \ldots, V_i(k_i))$. The scalar value $V_i(j), i \in [1, l], j \in [1, k_i]$ denotes the probability of generating the value j for gene i.

Initially (see point 2 in figure 1), and without any information, the $V_i(j)$ are initialized to $\frac{1}{k_i}$. This corresponds to a uniform distribution where the gene values are all equiprobable, which is exactly what a standard GA usually does in a finite population.

Then (see point 3 in figure 1), 12 individuals are generated and displayed in order to let the user select the style sheets that look the best. This is performed by applying the generated style sheets to the pages that the user has initially provided, and by displaying the obtained HTML pages. The user may select individuals, unselect the previously selected individuals, zoom on a given style sheet. He or she may also edit an individual in order to modify its genes directly. These modifications are directly taken into account in the genetic representation of the individual, and will be simply considered as mutations.

Let $I_1, \ldots, I_m$ denote the individual which have been selected by the user. These individuals are thus supposed to contain more interesting characteristics/genes than the others, and this is why these individuals are used to update the vectors $(V_1, \ldots, V_l)$. For this purpose, let us consider the vectors $(V'_1, \ldots, V'_l)$ but computed using the selected individuals $I_1, \ldots, I_m$ only. The following equation is simply going to move $(V_1, \ldots, V_l)$ closer to $(V'_1, \ldots, V'_l)$:

$$\forall i \in [1, l], V_i \leftarrow (1 - \alpha_{k_i})V_i + \alpha_{k_i}V'_i \tag{1}$$

where α_{k_i} are scalars between 0 and 1 and which values are detailed in the following. This equation can also be rewritten in the following way for each vector component:

$$\forall i \in [1, l], \forall j \in [1, k_i], V_i(j) \leftarrow (1 - \alpha_{k_i})V_i(j) + \alpha_{k_i}V'_i(j) \tag{2}$$

Then, mutation aims at moving $(V_1, \ldots, V_l)$ closer to uniformity, and is thus performed on each vector component in the following way:

$$\forall i \in [1, l], \forall j \in [1, k_i], V_i(j) \leftarrow (1 - p_{mut})V_i(j) + p_{mut}\frac{1}{k_i} \tag{3}$$

1. The user possibly **updates** the database, and **provides** the algorithm with a set of HTML pages,
2. **Initialize** uniformly the l probability vectors $V_1, \ldots, V_l$:

$$\forall i \in [1, l], \forall j \in [1, k_i], V_i(j) = \frac{1}{k_i}$$

3. **Generate** randomly the individuals according to $V_1, \ldots, V_l$. Initially, 12 individuals are generated, but as soon as a selection has taken place, selected individuals are kept in the population while the other non selected individuals will be removed and replaced by the generated offspring,
4. **Display** the 12 individuals. "Parents" individuals (that were selected by the user in the previous generation and that have been kept in the current population) are displayed at the same place on the screen, and are marked as selected. The visualization is obtained by applying the style sheets to the user initial HTML pages,
5. The user **selects** the individuals that he or she favors, or **unselects** previously selected individuals that he or she now dislikes. The user may also **adjust** the parameters p_{mut} and t in an intuitive way. Furthermore, he or she may directly **edit** an individual and **modify** its genes, and the user may also **save** an individual,
6. **If** at least one individual has been selected **Then:**
   - **Compute** the proportions $(V_1', \ldots, V_l')$ of gene values among the selected individuals only,
   - **Update** $(V_1, \ldots, V_l)$:

$$\forall i \in [1, l], \forall j \in [1, k_i], V_i(j) \leftarrow (1 - \alpha_{k_i})V_i(j) + \alpha_{k_i}V_i'(j)$$

7. **Mutate** $(V_1, \ldots, V_l)$:

$$\forall i \in [1, l], \forall j \in [1, k_i], V_i(j) \leftarrow (1 - p_{mut})V_i(j) + p_{mut}\frac{1}{k_i}$$

8. **Go to 3** or **Stop.**

Fig. 1. The general algorithm that generates style sheets (see additional explanations in text).

One should notice that if the user does not select any individual, which means that he or she is satisfied by none of the generated individuals, then as a consequence the IGA performs mutation only. The IGA can thus hopefully escape from local minima in the search space.

Once the new vectors V_i have been computed, the algorithm can generate the next generation. One should notice that all selected individuals in the previous generation are not modified, and these individuals are furthermore displayed at the same location on the screen and marked as "selected". The reason for keeping these individuals is that one should not loose a "good" individual, because in this case the user may have the feeling that his or her preferences are not taken into account. These individuals are displayed at the same location in order to

avoid to surprise the user by changing the position of individual on the screen.
Our GA is thus elitist because it only replaces unselected individuals.

Then a new cycle can start, and this lasts until the user is satisfied by the
proposed style sheets.

One interesting feature of IGAs is that the user may change his or her mind
during the search and may drive the algorithm to another area in the search
space at any time. The notion of a "good" style sheet may thus evolve over
time, and the ability of GAs to track a moving optimum is thus extremely useful
for the user.

The values of α_{k_i} are computed in the following way:

$$\alpha_{k_i} = 1 - e^{\frac{1}{t}ln(\frac{\epsilon}{1-\frac{1}{k_i}})} \tag{4}$$

A full explanation of this formula is beyond the scope of this paper, but we
can give an intuitive explanation here: let us consider that one characteristic j for
gene i is selected at each generation. If one wants to efficiently take into account
the user preferences, which implies that $V_i(j)$ converges from $\frac{1}{k_i}$ (uniformity)
to a probability $1 - \epsilon$ (as close to 1 as desired, assuming that the effects of
mutation are negligible, i.e. $p_{mut} << \epsilon$) in a given number of generation t, then
it is possible to demonstrate that α_{k_i} must be set to the value mentioned in the
previous formula. In the next section, we show that the user can modify t in a
very intuitive way in order to obtain a quick or slow convergence, because this
amounts to take into account its selections quickly or slowly. In this paper, we
have set $\epsilon = 1 - 0.5^{\frac{1}{t}}$, which ensures that the style sheet with the interesting
characteristics will be generated with a probability of 0.5 after t generations.

3.4   Interactions with the User

One of our main goal is to provide the user with a very intuitive and ergonomic
tool which can be used by anybody and not especially computer scientists. As
will be shown in the next section, the IGA contributes to this goal by letting the
user select individuals based on their graphical representation. We add in this
paper at least two other points in this context.

Setting the parameters of a GA can be difficult, even for a computer scientist.
Here it is very important to let the user adjust the mutation rate p_{mut} and the
time to convergence t. We have renamed those parameters in a more explicit way.
p_{mut} is called "Diversity" parameter and may be dynamically adjusted during
the search with a simple ruler. In the same way, t has been named "Importance of
choices" and may also be adjusted. Increasing the "diversity" parameter, i.e. the
mutation rate effectively has the effects that the user expects: new genes appear
in the individuals which thus look more "diverse". Increasing the "Importance
of choices" fastly drives the genetic search toward the selected individuals. In
this way, these parameters are more explicit from the user point of view even for
someone who has no knowledge of GAs and related techniques.

Sometimes the IGA can find a style sheet which is very close to the optimum,
but not exactly optimal. For instance, the "bullet" gene may not have the desired

value, while all the other elements have been well determined. Obtaining full converge of the "bullet" gene toward a value that satisfies the user could be long. This is the reason why we have added the possibility to directly edit the displayed individuals. In this way, the user may set a given value for a given gene at any generation. This modification is taken into account by the algorithm as if it had been performed by a genetic operator like mutation for instance. It can greatly speed up the time to convergence and sometimes provide a great help to the user: Imagine can be really viewed as a fully interactive style sheet editor which may suggest choices to the user or may take into account more direct user commands.

4   Results

A first series of results consists in optimizing an artificial HTML page which contains all the genes considered at the beginning of this paper (which includes rules, bullets, titles, etc). This initial page is represented in figure 2. One can notice that many style elements must be determined in order to get the final "good looking" page.

At the beginning of the run, we have selected all possible values in the database (see tables 1 and 2). Then the algorithm proposes an initial generation

Fig. 2. Here is represented an example of an initial HTML page which is going to be optimized with Imagine.

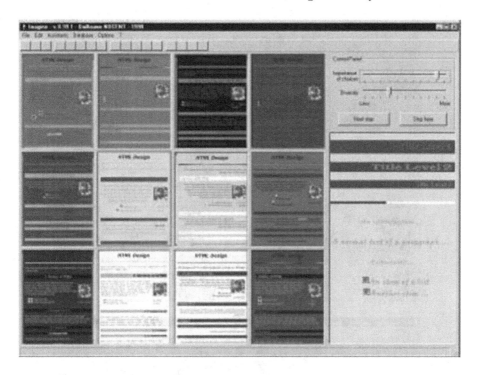

Fig. 3. A view of Imagine main screen. This screen is presented to the user who may select/unselect individuals, zoom on a given individual, navigate through his or her set of pages or web site (to which the generated style sheet has been applied), edit an individual, save an individual, etc.

of individuals, and the selection/generation cycles can start. Figure 3 gives an examples of Imagine main screen. One can notice that 12 individuals are represented. These representations are obtained by applying the style sheets to the user initial page. As mentioned in figure 3, the user has the possibility to perform many actions. For instance, with a double-click on an individual, the user may zoom on this individual. On the upper-right corner of the screen, the user may adjust the parameters. On the bottom-right corner of the screen, Imagine dynamically displays the main characteristics of the style sheet which is beneath the mouse cursor. This is where the individuals can also be edited. The user can select an individual by clicking on it.

We have obtained many different kind of pages like those presented in figure 4. The possibilities of Imagine are very large. If one considers the actual set of genes and all their possible values, one may obtain the size of the search space by multiplying the number of values k_i of every gene i. One obtains this way a search space of 10^{28} possible style sheets. This search space can be enlarged or reduced when the user modifies the database.

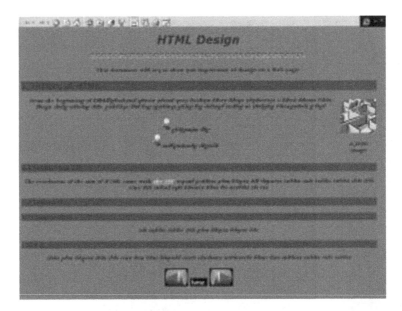

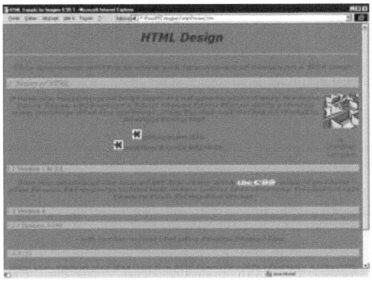

Fig. 4. Two examples of style sheets which can be generated by Imagine (in this example, we have not used the "editing" abilities of Imagine, so this is strictly the results obtained with the IGA in a dozen of generations). The reader may find more examples at the following URL: http://www.rfai.li.univ-tours.fr/webrfai/chercheurs/venturini/imagine.htm.

We have applied Imagine to a whole set of pages which can be viewed at the following URL: http://www.rfai.li.univ-tours.fr/webrfai/chercheurs/venturini/imagine.htm. In this case, these pages have been initially designed with LaTex, then they have been converted to HTML. Then Imagine has been used to generate style sheets for these pages. It is important to notice that Imagine can easily handle any kind of HTML pages (provided that the HTML syntax is correct), so the user may use initially any HTML editor to define the content of his or her pages.

Finally, we are currently starting an industrial collaboration with the Alderan Company. Alderan provides many internet-based services to its customers and Imagine is being implemented and added to these services. It will thus be soon used in real world applications.

5   Conclusion

We have presented in this paper a new method for the interactive generation of HTML style sheets. The aim of this method is to improve and optimize the look of a web site. The originality of this method comes from the fact that it can take into account the user preferences in a very intuitive and simple way. An IGA is used to efficiently exploit the information contained in the user selected individuals. This IGA uses the gene frequencies to generate new individuals, a technique which, as far as we know, has not been used yet in IGAs studies. This prototype has been applied to several artificial cases and is being currently implemented for real-world applications.

This work can be extended in several ways. For instance, we are actually studying how to "filter" individuals and to eliminate them without any presentation to the user. Some individuals are obviously uninteresting (for instance, when the background color is not compatible with the text color). This deletion could be done by adding some constraints in the genetic search. These constraints could be encoded for instance as a matrix of compatibility between colors. In the future, we also argue that IGA can be used more generally to generate other kind of computer interfaces like in software engineering (Seers 1993) and certainly have a role to play in human-computer interactions.

References

Baluja S. (1994). "Population-based Incremental Learning: A Method for Integrating Genetic Search Based Function Optimization and Competitive Learning". Tech. Rep. CMU-CS-94-163, Carnegie Mellon University.

Caldwell C. and Johnston V.S. (1991), Tracking a criminal suspect through "face-space" with a genetic algorithm, Proceedings of the Fourth International Conference on Genetic Algorithms, 1991, R.K. Belew and L.B. Booker (Eds), Morgan Kaufmann, pp 416-421.

Dawkins R. (1986), *The Blind Watchmaker*, Longman, Harlow, 1986.

Graf J. and Banzhaf W. (1996), Interactive evolution in the framework of simulated natural evolution, Evolution Artificielle 95, E. Lutton, E. Ronald, M. Schoenauer and D. Snyers (Eds), Lecture Notes in Computer Science 1063, Springer Verlag, pp 259-272.

Holland J.H. (1975). *Adaptation in natural and artificial systems*. Ann Arbor: University of Michigan Press.

Sears A. (1993), Layout appropriateness: a metric for evaluating user interface widget layout, *IEEE Trans. on Software Engineering*, Vol. 19, No 7, July 1993, pp 707-719.

Sebag M. and Ducoulombier A. (1998). "Extending Population-Based Incremental Learning to Continuous Spaces". Proceedings of PPSN'98, Parallel Problem Solving fron Nature, T. Baeck and M. Schoenauer Eds, Springer-Verlag, 1998.

Sims K. (1991), Interactive evolution of dynamical systems, Proceedings of the first European Conference on Artificial Life 1991, F.J. Varela and P. Bourgine (Eds), MIT press/Bradford Books, pp 171-178.

Smith J.R. (1991), Designing biomorphs with an interactive genetic algorithm, Proceedings of the Fourth International Conference on Genetic Algorithms, 1991, R.K. Belew and L.B. Booker (Eds), Morgan Kaufmann,, pp 535-538.

Syswerda G. (1993), Simulated crossover in genetic algorithms, Proceedings of the second workshop on Foundations of Genetic Algorithms, L.D. Whitley (Ed.), Morgan Kaufmann, pp 239-255.

Todd S.P. and Latham W. (1991), *Mutator, a Subjective Human Interface for Evolution of Computer Sculptures*, IBM United Kingdom Scientific Center Report, 1991.

Venturini G. and Montalibet D. (1995), Analyse exploratoire d'une base de données utilisant un algorithme génétique interactif et la programmation génétique, Journées de la Société Francophone de Classification (SFC), Namur, Belgique.

Venturini G., Slimane M., Morin F. and Asselin de Beauville J.-P. (1997), On Using Interactive Genetic Algorithms for Knowledge Discovery in Databases, Proceedings of the seventh International Conference on Genetic Algorithms, 1997, T. Baeck (Ed.), Morgan Kaufmann, pp 696-703.

Problem-Specific Representations for Heterogeneous Materials Design

Alain Ratle

Département de génie mécanique, Université de Sherbrooke,
Sherbrooke, Québec, J1K 2R1 Canada
(current address: Laboratoire de mécanique des solides,
Ecole Polytechnique, 91128 Palaiseau, France)
ratle@lms.polytechnique.fr

Abstract. This paper investigates the use of problem-specific data structures and operators in evolutionary optimization for a specific class of combinatorial design problems. The problem consists of finding the optimal distribution of two or more phases of a sound absorbing material on a three-dimensional network, in order to maximize sound absorption properties. The natural structure of the problem is by the way very far from the linear chains classically used by evolutionary algorithms (EAs). Special operators exploiting the three-dimensional structure are proposed and compared with other operators that are working on a linear chain representation. The formers are potentially useful since the natural neighborhood relationships are lost in a linear representation.

1 Introduction

EAs have always relied on a linear chain encoding of the problem to be solved. The initial idea was that following schema theorem [Hol75], some structures and operators were more likely to give better results than others, regardless of the problem at hand. This was inspired from the fact that nature has evolved complex and efficient structures using a chain-like genetic code [Bäc96]. However, since most real-world problems are very different from nature's problems, there is no reason to believe that nature's solutions should give optimal results for artificial problems. Many of the real-world optimization problems do not fit easily into a linear representation, and better results have often been obtained using problem-specific data structures and operators. An example is the traveling salesman problem for which various data structures and operators have been tried out [FM91]. Moreover, recent works have suggested that any "blind" change of representation is futile as long as the new coding is not correlated with the problem at hand [Cul98]. This follows from the No Free Lunch theorem [WM97] which states that no optimization algorithm should be expected to give better results on average than any other, unless it has some correlation with a specific problem.

This paper presents the case of a multiphased material design problem which is naturally formulated into a three-dimensional structure. The objective is to

C. Fonlupt et al. (Eds.): AE'99, LNCS 1829, pp. 111–122, 2000.
© Springer-Verlag Berlin Heidelberg 2000

find out the optimal distribution of two or more types of sound absorbing material elements in order to maximize the absorption properties. The problem is subject to constraints on the desired fraction of each phase. In a first approach, evolutionary operators preserving the constraints are developed. These operators are however designed with no regard to the particular data structure. In a second approach, evolutionary operators based on the natural structure are elaborated, and the optimization results are compared with the former approach.

2   The Physical Problem

Porous materials are often used for various noise control applications [All93]. It have been pointed out by Allard [All93] that the efficiency of these materials might be improved by an heterogeneous stratification. More recently, Atalla et al. [APA96] have suggested that the use of three-dimensionally heterogeneous networks might yields even better performances, but this hypothesis has not been tested out.

The aim of this paper is to give an answer to the question as whether or not an optimal 3-D distribution of multiple phases can give better solutions compared to a simple one-dimensional stratification. The proposed approach is to use problem-specific EAs as an experimental tool to gain knowledge on this question. Recent results by Ratle & Atalla [RA98] have shown that constraint preserving operators greatly ease the search by a restriction of the search space. However, the exploitation of the natural data structure in the design of evolutionary operators has never been addressed.

In the present case, the material is modeled by a finite element mesh having $N_x \times N_y \times N_z$ elements, and made of two or more different materials. The resulting multiphased material is placed at one end of a semi-infinite acoustical wave guide, as shown on Figure 1. The boundaries of the domain are assumed to be rigid and the porous medium is excited by a plane wave. The resolution gives the absorption coefficient at a specific frequency, and the optimization criteria is the average value of this coefficient in some frequency range.

The design efficiency relies on the availability of a suitable optimization method. A major constraint is that only a small number of fitness function evaluations can be allowed, due to computational cost. The chosen approach [RA98] transforms a parametric problem into a combinatorial "N choose $m_1, m_2, \ldots, m_M$" problem, where N is the total number of elements and the m_i's are the number of elements required for each of the M materials. The coding consists of a sequence of integers where each one is mapped to a type of material. In the binary case, a 0 represents an element of the base material, and a 1, an element of the so-called additive material. Such a representation have often been used for topological design of mechanical shapes (see for example Kane and Schoenauer [KS95].

Elementary combinatorics shows that the number of possible solutions with N elements and M materials is equal to $N!/\prod_{i=1}^{M}(m_i!)$. The optimization problem is worked out with respect to two concepts. First, problem-specific ope-

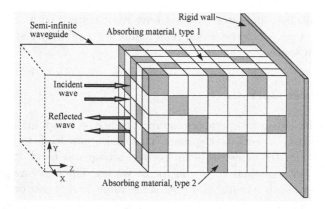

Fig. 1. Heterogeneous porous material made up from a network of homogeneous patches set on a rigid impervious backing and coupled to a semi-infinite waveguide.

rators for implicit constraint preservation are developed. The second concept consists of considering the natural data structure in operators design. This concept should help to answer the two following questions:

1. Is it better to preserve the whole 3-D structure, or only the average distribution along the principal direction?
2. Is it better to use a local search operator which is defined in the problem space, or one that is defined in the representation space, the latter being potentially highly disruptive in the problem space?

3   Evolutionary Operators for Specific Representations

3.1   Initialization Operator

Initializing a genetic code for the problem at hand requires a random string of length N containing exactly m_j characters of each category j, where $j = 1 \ldots M$. The proposed method is described as follow:

Begin initialization
    Initialize M counters to zero
    For $i = 1$ **to** N
      Do
          $j \leftarrow$ integer random value $\in \{1, \ldots M\}$
          If counter[j] $< m_j$ (feasible value found)
              Increment counter[j]
              (genetic code[i]) $\leftarrow j$
          End if
      While feasible value not found
    End for
End initialization

The initialization operator does not have to be defined with consideration of the data structure, since the initial individuals are randomly distributed in the representation space.

3.2   Crossover Operators

The design of a crossover operator requires a choice on which informations must be preserved from parents to offsprings. This choice may leads to radically different operators for the same problem. For the problem at hand, two approaches are proposed with implicit constraint preservation. The first one preserves the positions where both parents share the same value, and the second one tries to preserves the longest sub-solution common to both parents. A third operator designed for one-dimensional information preservation is finally presented.

Identical Points Preservation Crossover. A first approach consists of preserving all the positions in the coding where both parents have the same material. For the other positions, any value is acceptable, as long as the constraints are respected. This procedure is termed Identical Points Preservation (IPP) crossover. Given two parents a and b, the offspring c is created by the following procedure:

Begin IPP crossover
    Initialize M counters to zero
    For $i = 1$ **to** N
        If $(a_i = b_i)$: $c_i \leftarrow a_i$ and Increment counter$[a_i]$
    End for
    For $i = 1$ **to** N
        If $(a_i \neq b_i)$
            Case (counter$[a_i] = m_{a_i}$ **and** counter$[b_i] < m_{b_i}$) : $c_i \leftarrow b_i$
            Case (counter$[a_i] < m_{a_i}$ **and** counter$[b_i] = m_{b_i}$) : $c_i \leftarrow a_i$
            Case (counter$[a_i] < m_{a_i}$ **and** counter$[b_i] < m_{b_i}$) : select $c_i \in \{a_i, b_i\}$
            Case (counter$[a_i] = m_{a_i}$ **and** counter$[b_i] = m_{b_i}$) : select $c_i \notin \{a_i, b_i\}$
            Increment counter$[c_i]$
        End if
    End for
End IPP crossover

This operator introduces a high-level of randomization whenever few common building blocks are present between the two parents. This randomization vanishes when the parents become very similar. Because the IPP crossover preserves all the stable elementary positions, it is by the way a structure-preserving operators, since large blocks of common elements between both parents will be preserved. This principle, characteristics common to both parents are passed to the children, have been previously stated by Radcliffe [Rad91], and by Surry and Radcliffe [SR96] as the Random Respectful Recombination (RRR).

Longest Common Substring (LCS) Crossover. The second crossover approach is the preservation of an emerging partial ordering common to both parents, that is, a substring of length $\ell \leq N$. The proposed algorithm is described as follow:

1. Find the longest common substring between the two parents;
2. copy this substring into the offspring at the position indicated by the parent with highest fitness;
3. fill in the remaining positions with the variables given by the second parent in their relative order.

Optimal Crossover Strategy. A better approach is the design of an algorithm that selects the optimal operator between IPP and LCS in every situation. Since a common substring between two parents is likely to be significant only if it is longer than a certain threshold, an optimal crossover strategy would be to use LCS crossover if a long enough substring is found, and IPP crossover otherwise.

Layer crossover. The layer crossover ensures the preservation of only the average composition of each layer along the z axis, regardless of the value carried by the individual elements. This crossover is performed as follow, for the binary case:

1. Calculate for the two parents the number of additive elements $N_{add}(k)$ on each layer $k = 0, \ldots, N_z - 1$.
2. Perform crossover over the $N_{add}(k)$'s from the two parents using binary selection between parent 1 and parent 2, i.e. choose either one or the other.
3. Repair the offspring in order to preserve the total number of additive elements: while $\sum_k N_{add}(k) > t$, pick a layer k randomly and decrease $N_{add}(k)$ by one, or do otherwise (increase it) if $\sum_k N_{add}(k) < t$.
4. Select randomly $N_{add}(k)$ positions on each layer k of the offspring and turn them to 1, turn the others to 0.

3.3 Mutation Operators

Any mutation operator that consist of a permutation of elements ensures constraint preservation for this problem. Three operators are suggested:

1. The *point* operator, which consists of choosing two random elements (points), and swapping their values. Since both elements may have the same value, a verification is done to avoid redundant solutions.
2. The *displacement* operator consists of choosing a substring between two randomly chosen points, cutting this string and reinserting it elsewhere.
3. The *inversion* operator consists of choosing a substring between two randomly chosen points, and inverting the order of the values in this substring.

The first of these operator works on a one-dimensional structure, and is therefore the most simple move that can be defined. The two other operators act on a two-dimensional representation which might be correlated or not to the problem-space.

Another mutation operator which works directly on the three-dimensional problem space is also proposed. This operator, the block mutation, produces an inversion of a 3-D block in the problem space. Two corners of a block are first selected (two triplets (i, j, k)), and a mirroring direction is chosen among the three possibilities. Elements into the block are then swapped pairwise in the selected direction. This operator does not breaks down heavily the natural data structure, compared to the chain mutation operator. Once again, null operations are detected in order to avoid useless reevaluations. A null operation arise whenever the block has a unit or null thickness in the mirroring direction.

4   Numerical Results

Computational experiments have been performed on 3 problems of various size. The first two cases address only the one-dimensional heterogeneity with 15 and 40 layers in the z direction. The first case is in a low frequency range, where high absorption values are very unlikely, and the second case in a mid-frequency range. The third case is a 3-D problem with 250 elements, $N_x = 5$, $N_y = 5$ and $N_z = 10$. Two materials are available with 200 elements of base and 50 elements of additive material. This case is also in a mid-frequency range. Since each evaluation requires the solution of a finite element problem, CPU time is a major limitation[1] and a maximum of 1000 evaluations have been allowed in each case. This represents the stopping criteria of the algorithm. The 3 cases are summarized on Table 1. For the cases 1 and 2, the optimal solution is known, and is compared with optimization results. However, for the case 3, the optimal solution is not known. In all the cases, results are averaged over five runs.

4.1   Constraint Preservation Operators

Mutation Operators. Constraint-preserving mutation operators have been evaluated with a simplified EA, with a population of size 2, binary tournament selection, no crossover, an elitist survival of the best solution in the current population, and a probability of mutation $p_m = 1$. Results are presented on Figure 2a to 2c for the 3 problems. In all the cases, the point operator gives

Table 1. Description of the three test cases

Case no	1	2	3
Material no	Number of elements		
1	5	20	200
2	5	20	50
3	5	0	0
Total	15	40	250
Number of solutions	7.57×10^5	1.38×10^{11}	1.35×10^{53}

---

[1] The resolution for $N = 250$ takes about 1 minute of CPU on a Pentium 400 MHz running under Linux 2.0, and the complexity of finite elements problems is $O(N^3)$.

better results than any other, but the difference is larger for the 250 variables problem. These differences are explained by representation issues. For the 15 and 40 variables cases, there is a perfect correlation between the problem space and the representation space. For the third case, the problem space is uncorrelated with the representation, and local search operators defined in one space are highly disruptive in the other. The point operator is not affected by this problem.

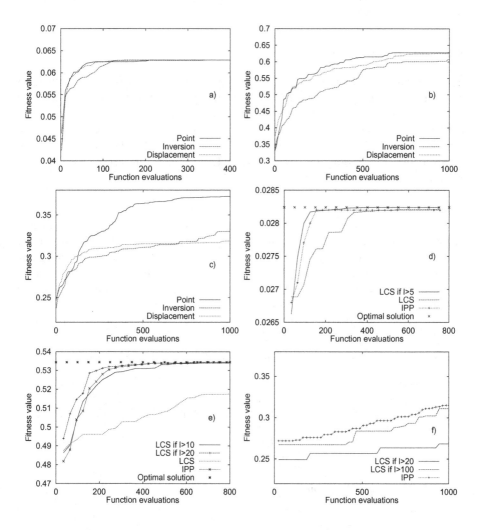

Fig. 2. Comparison of the three mutation operators, for a) 15 variables problem, b) 40 variables, and c) 250 variables. Comparison of the crossover strategies for d) 15 variables problem, e) 40 variables, and f) 250 variables.

Crossover Operators. The two constraint preservation operators have also been compared for the three cases, using an EA with a population size of 25, point mutation with $p_m = 1/2$, probability of crossover $p_c = 1$, binary tournament selection and the elitist survival of the best 5 individuals. Results are presented on Figure 2d for the case 1, and 2e for case 2. These two cases show that the preservation of common positions (IPP) is always better than the preservation of the greatest common substring (LCS) when considered alone. However, a slight improvement is obtained if the LCS operator is employed only if a chain having a certain threshold length is found. For case 3, results presented on Figure 2f shows that the IPP operator is always better. This can also be explained by the disruption problem.

4.2   Integration of Data Structures

This section considers only the 250 variables problem, since the effect of the data structure is present only for 3-D distributions. All the experiments are based on the same EA with a population size of 25, binary tournament selection, elitist survival of the best 5 solutions, a maximum number of 1000 function evaluations, and variable probabilities of crossover and mutation. In the first 4 series of experiments, the various crossover and mutation operators are compared. The next 4 series compare the effect of the rate of application of a same operator. The 8 test cases are described on Table 2. This table also gives the best fitness values obtained on average after 1000 evaluations.

An interesting result is that in all cases except case 7, the use of a crossover operator brings no improvement compared to the use of mutation alone. The best results are even obtained with the most simple operator, the point mutation. This suggests that improvements come mainly from small local perturbations rather than dramatic changes or recombinations. Comparing the two crossovers alone (case 2), it is observed that the preservation of average content of each layer gives better results than the preservation of individual positions

Table 2. Summary of cases settings and results ($p_m = 1$ in all cases except 2).

Case	Mutation	Crossover	p_c	Fitness	Case	Mutation	Crossover	p_c	Fitness
	Point		-	0.349				0.0	0.357
1	Chain	No	-	0.322	5	Point	Positions	0.5	0.347
	Block		-	0.336				1.0	0.316
2	No	Positions	1.0	0.282				0.0	0.357
		Layers	1.0	0.307	6	Point	Layers	0.5	0.328
								1.0	0.278
3	Point	Positions	1.0	0.285				0.0	0.329
		Layers	1.0	0.324	7	Block	Positions	0.5	0.343
								1.0	0.335
4	Block	Positions	1.0	0.325				0.0	0.329
		Layers	1.0	0.286	8	Block	Layers	0.5	0.309
								1.0	0.302

(IPP). This suggests that the IPP crossover alone can hardly introduce new and useful information without the help of a mutation operator. The hypothesis is confirmed by the observation of cases 5 to 8. In these cases, the use of a mutation operator together with the two crossovers gives the advantage to the position-preservation crossover, compared to the layer crossover. This means that there is some interest in preserving the 3-D distribution rather than only a one-dimensional distribution, as long as the evolutionary operators are able to introduce a sufficient amount of diversification.

Another aspect of the problem is the correlation between the various operators in the same algorithm. Case 7 is the only one where an improvement is brought in by the crossover operator. It is also the only one where both structure-preserving operators are employed. This suggests that care should be taken in matching the crossover and mutation operators, the best results are obtained when both are defined with respect to the same kind of information structure.

5 Physical Significance

A comparison of the solutions found by evolutionary optimization for the 250 variables problem with other types of solutions is given on Table 3. The best solution found in all the optimization runs limited to 1000 evaluations, and the average of the 10 best runs are compared with a classically known solution, which consists of stacking the materials in increasing order of fluid flow resistance. This is equivalent to placing all the 50 additive elements in the first two layers. It is observed from these data that the classical solution gives better results than a random distribution of the two phases. However, the solutions found by the EAs give significant improvements. It should be noted that although the limit of 1000 evaluations represents a reasonable computational cost, it does not guarantees optimality. Solutions found after 6000 evaluations show some improvement, but with a computational cost 6 times higher. The standard error decreases dramatically between 1000 and 6000 evaluations, thus indicating that the algorithm tends to reach some limiting value after 6000 evaluations.

The physical distribution of the additive elements is illustrated for two cases. Figure 3 shows a "good" solution with a fitness value of 0.343, and the best solution with a fitness of 0.380. The base material elements are represented on these two figures by the black lozenges, and the additive elements by the circles.

Table 3. Comparison of optimization results and standard solutions.

Description	Average	Standard error
Best solution found after 1000 evaluations	0.3647	-
Average of 10 best solutions after 1000 evaluations	0.3565	0.0048
Classical solution (separated phases)	0.3295	-
Random solution (average of 50)	0.2339	0.0165
Best random solution among 50	0.2710	-
Best solution found after 6000 evaluations	0.3803	-
Average of 10 best solutions after 6000 évaluations	0.3795	0.0007

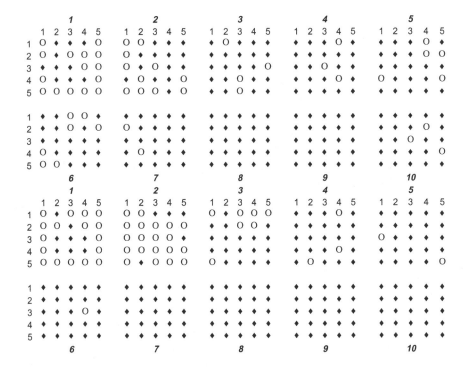

Fig. 3. Distribution of additive elements for a moderately good solution (fitness=0.343) on top, and the best solution found (fitness=0.380) on bottom.

The layer 1 corresponds to the (leftmost) incidence layer, and the layer 10 is backed by the rigid wall. Comparing these cases, it seems that the more the additive elements are pushed to the left, the better is the solution. The best case would be with all the additive elements in the first two layers. However, data on Table 3 show that this is not true. There must exist a non-trivial solution which maximizes the sound absorption coefficient, and which cannot be manually extrapolated from past experience.

6   Discussion and Conclusions

This paper presented an original design problem which can hardly be dealt with using general-purpose algorithms. Even though the case is naturally formulated in a binary, or low-cardinality characters string, classical genetic algorithms are of limited use, due to the constraints and the particular data structure. The comparison of the constraint-preserving mutation operators have shown that in all the cases, an operator working on single elements gives better results than operators working on longer strings. This effect is clearly more pronounced for three-dimensional cases, due to the disruption between problem space and representation space. The same effect has been observed for crossover operators.

In all the cases, but more significantly with a 3-D distribution, preserving the individual positions is better than preserving a substring.

Evolutionary operators have been developed with respect to the specific data structure of the problem. A mutation operator acting on a block of elements defined in the problem space gives better results than one working on a string in the representation space. However, the best results are still obtained using a mutation working on a pair of single elements. In the same way, two crossover operators have been proposed, the first one preserving the individual positions where both parents share the same value, thus preserving their spatial distribution, and the second preserving only the average content of each layer in the main direction. Results have shown that the preservation of the 3-D structure has a positive effect, as long as a sufficient amount of new and relevant information is introduced by a suitable mutation operator. These results also suggests that the correlation between operators employed in the same EA has a great relevance.

The global results show that evolutionary optimization brings significant improvements in the quality of solutions compared to the solution classically known by practitioners and also compared to random solutions, in spite of the fact that the allowed computational resources are not sufficient to ensures optimality. This point enlightens an interesting feature of EAs: even when no guarantee of optimality can be stated, evolutionary optimization often give truly interesting solutions from an engineering point of view, compared to usual practices.

It is also interesting to note that mixing up more than one crossover operator has given in some cases better results than the use of any operator alone. This suggests that a possible improvement would be to use a self-adaptive algorithm, where many different crossover and mutation operators are available, since they all have good and bad points. The probability of use of each one can be coded in the individuals, in a similar way as mutation amplitudes are coded in the self-adaptive evolution strategies [Bäc96], with the updating of the probabilities based on previous performances.

References

All93.   J.F. Allard. *Propagation of sound in Porous Media: Modeling sound absorbing materials*. Chapman & Hall, London, 1993.

APA96.   N. Atalla, R. Panneton, and J.F. Allard. Sound absorption by non homogeneous thin porous layers. *Acta Acoustica*, 83:891–896, 1996.

Bäc96.   T. Bäck. *Evolutionary Algorithms in Theory and Practice*. Oxford University Press, 1996.

Cul98.   J.C. Culberson. On the futility of blind search: an algorithmic view of "no free lunch". *Evolutionary Computation*, 6(2):109–127, 1998.

FM91.   B.R. Fox and M.B. McMahon. Genetic operators for sequencing problems. In G.J.E. Rawlins, editor, *Foundations of Genetic Algorithms*, pages 284–300, 1991.

Hol75.   J.H. Holland. *Adaptation in Natural and Artificial Systems*. MIT Press, 1975.

KS95.   C. Kane and M. Schoenauer. Genetic operators for two-dimensional shape optimization. In J.-M. Alliot, E. Lutton, E. Ronald, M. Schoenauer, and D. Snyers, editors, *Artificial Evolution*, 1995.

RA98.   A. Ratle and N. Atalla. Evolutionary optimization strategies for the combinatorial design of heterogeneous materials. *J. Evolutionary Optimization*, 1(1):77–88, 1998.

Rad91.   N.J. Radcliffe. Forma analysis and random respectful recombination. In R.K. Belew and L.B. Booker, editors, *Proceedings of the First International Conference on Genetic Algorithms*, pages 222–229, 1991.

SR96.   P.D. Surry and N.J. Radcliffe. Formal algorithms + formal representations = search strategies. In H.-M. Voigt, W. Ebeling, I. Rechenberg, and H.-P. Schwefel, editors, *Parallel Problem Solving from Nature IV*, pages 366–375, 1996.

WM97.   D.H. Wolpert and W.G. MacReady. No-free-lunch theorems for optimization. *IEEE Transactions on Evolutionary Computation*, 1(1):67–82, 1997.

A Hybrid Evolution Strategy for Mixed Discrete Continuous Constrained Problems

Laurence Moreau-Giraud and Pascal Lafon

Laboratoire des Systèmes Mécaniques et d'Ingénierie Simultanée (LASMIS)
Université de Technologie de Troyes
12 rue Marie Curie - BP 2060 - 10010 Troyes Cedex
Laurence.Giraud@univ-troyes.fr, Pascal.Lafon@univ-troyes.fr

Abstract. In this paper, a hybrid evolution strategy is proposed to solve mixed discrete continuous constrained problems. We consider that the functions of the problems are differentiable with respect to the continuous variables but are not with respect to the discrete ones. Evolutionary algorithms are well suited to solve these difficult optimization problems but the number of evaluations is generally very high. The presented hybrid method combines the advantages of evolutionary algorithms for the discrete variables and those of classical gradient-based methods for the continuous variables in order to accelerate the search. The algorithm is based on a dual formulation of the optimization problem. The efficiency of the method is demonstrated through an application to two complex mechanical design problems with mixed-discrete variables.

1 Introduction

Evolutionary algorithms have started to receive significant attention during the last decade [12,1]. Part of their success is due to the large domain of application of those methods, their robustness and their gain of flexibility. Those methods are specially well suited for solving difficult optimization problems [4]. However, their computational cost is generally very high. A large number of evaluations must generally be performed for a satisfying result to be found.

On the other hand, deterministic methods, which exploit local information like gradient information are rapid methods when the gradients of the functions can be calculated. But their domain of application is reduced to the class of differentiable problems.

Two-stage procedures of first using evolutionary algorithms to overcome multiple minima, followed by more traditional optimization methods to accelerate local convergence have been presented in the last decade [10,13]. In [9], we have proposed a two-stage procedure for the optimal design of mechanical systems, where the evolution strategy is used first during a fixed number of evaluations and the augmented Lagrangian method is applied in a second stage on the continuous

C. Fonlupt et al. (Eds.): AE'99, LNCS 1829, pp. 123-135, 2000.
© Springer-Verlag Berlin Heidelberg 2000

variables. But the major difficulty with these hybrid methods is to find the moment to stop the stochastic method to start the deterministic one.

The aim of this paper is to propose an original coupling of an heuristic method with deterministic ones for mixed discrete continuous constrained problems, which uses in parallel the advantages of the deterministic methods for the evolution of the continuous variables and the advantages of the evolutionary algorithms for the discrete variables. The hybrid method, proposed in this paper is based on the generalized evolution strategy (GES) developed for mixed discrete-continuous variables [2] and the augmented Lagrangian method [11,8]. The augmented Lagrangian function is used to evaluate each individual of the population and an update of the Lagrange multipliers is realized during the optimization process. But the main difference with GES is the mean of calculating the continuous variables of each offspring individual. Instead of determining them by recombination and mutation, the hybrid method realizes in the continuous space a step along a Lagrangian quasi-Newton search direction.

We show in the application of this hybrid method on two mechanical design problems that the incorporation of deterministic rules to determine the continuous variables of each offspring individual has allowed to accelerate the convergence significantly.

2 The Mixed Discrete Continuous Constrained Problem

The general mixed discrete continuous problem is defined as:
Minimize

$$f(x) \tag{1}$$

Subject to

$$g_j(x) \leq 0, j=1, ..., m \tag{2}$$

$$g_j(x) = 0, j=m+1,..., m+l \tag{3}$$

$$u \leq x \leq v, x \in \mathbf{R}^N = \mathbf{R}^{N_C} \cup \mathbf{R}^{N_D} \tag{4}$$

where f is the objective function and g_j are nonlinear inequality and equality constraint functions. The components of the mixed variable vector $x = (x_C, x_D)^T$ are divided into n_C continuous variables, expressed as $x_C \in \mathbf{R}^{N_C}$ and n_D discrete variables, expressed as $x_D \in \mathbf{E}^{N_D} \subset \mathbf{R}^{N_D}$ Here $\mathbf{E}^{N_D} \subset \mathbf{R}^{N_D}$ is the space of discrete values which are acceptable by the problem and can be randomly distributed.

In this class of problems, the first order derivatives of the objective function and constraints exist with respect to the continuous variables. The existence of discrete variables may arise from commercially available material sizes, standardization of parts... and the derivatives of the functions with respect of these variables can not always be computed. In this paper, we consider a class of problems for which the first derivatives of the functions with respect to the discrete variables can not be determined.

3 Hybrid Method

The combination of the discrete variables together with the nonlinear inequality and equality constraints results in complex optimization problems which are difficult to solve with classical gradient based methods. Evolution strategies, members of the class of evolutionary algorithms are well-suited to solve these difficult problems. But the number of evaluations necessary to find the exact solution for heavily constrained optimization problems is generally high.

Evolution strategies are based on the principle of evolution, i.e. survival of the fittest. Unlike classical methods, they do not use a single search point but a population of points called individuals. Each individual represents a potential solution to the problem. At each iteration, called a generation, a new population of χ offspring individuals is created by means of recombination and mutation, starting from the old population of μ parents ($\chi > \mu$). The μ best individuals out of the population of parents and offspring are selected to reproduce and replace the old population of parents. By means of these randomized processes, the population evolves toward better and better regions of the search space.

Deterministic methods assume that the objective and constraint functions are differentiable. They do not use probabilistic rules and they generally converge quickly. Those methods have a single search point. At each iteration k, a new point x^{k+1} is determined by the point x^k and a step in a calculated direction d^k :

$$x^{k+1} = x^k + \alpha^k \ d^k \tag{5}$$

The calculation of the step α^k and the direction d^k depend on the chosen deterministic method. They are generally based on the gradients of the functions.

As the gradients of the functions with respect of the continuous variables are available in the class of problems that we consider, we have used this information and combined the advantages of the evolution strategy for the evolution of the discrete variables and those of deterministic methods for the evolution of the continuous variables in a hybrid method.

The hybrid method uses a population of μ parent individuals $x^P = ((x_C^P, x_D^P),$ $p=1, \ldots, \mu)$ and χ offspring individuals $((x_C^s,", x_D^s,"), s=1,\ldots, \chi)$.

It is performed using an augmented Lagrangian function (see [11,8]). The augmented Lagrangian function is an association of a classic dual function and a quadratic penalty function. Its expression is as follows [11]:

$$L(x, \lambda, r) = f(x) + \sum_{j=1}^{m} \left(\lambda_j \psi_j(x) + r.\psi_j^2(x) \right) + \sum_{k=m+1}^{m+l} \left(\lambda_k g_k(x) + r.g_k^2(x) \right) \tag{6}$$

$$\text{with: } \psi_j(x) = Max\left\{ g_j(x), -\frac{\lambda_j}{2r} \right\} \tag{7}$$

In the hybrid method, iterations are realized between the space of primal variables (x) and the space of dual variables (λ, r).

The algorithm below shows the different steps of the hybrid method:

Step1: Initialisation
    Step1.1: $t:=0$
    Step1.2: Set Lagrange multipliers $\lambda_j^t = 0, j=1, \ldots, (m+l)$
    Step1.3: Set a penalty parameter $r^t = r_0$
    Step1.4: Initialize the population P
Step2: Evaluation of P
Step3: While not terminate do:
    Step3.1: Initialize the Hessian matrix of the augmented Lagrangian function, associated to each individual in the population P
    Step3.2: While not terminate do
        a) $(x_D^s{}'', p^s{}'')$ = recombination and mutation of the discrete variables and strategy parameters for the offspring individual s, $s \in [1, \chi]$
        b) $x_C^s{}''$ = computation of the continuous variables of the offspring individual s with deterministic methods, $s \in [1, \chi]$
        Details in procedure A
        c) Evaluation of P'', population of χ offspring individuals $[(x_C^s{}'', x_D^s{}'', p^s{}''), s=1, \ldots, \chi]$
        d) P: selection $(P'' \cup P)$
    Step3.3: Update of the Lagrange multipliers by:

$$\lambda_j^{t+1} = \lambda_j^t + 2r^t Max\left\{ g_j(x^b), -\frac{\lambda_j^t}{2r^t} \right\} \qquad j = 1, \ldots, m \tag{8}$$

$$\lambda_j^{t+1} = \lambda_j^t + 2r^t g_j(x^b), j = m+1, \ldots, m+l \tag{9}$$

x^b : best solution among the $(\mu + \chi)$ individuals
    Step3.4: Update of the penalty parameter

The remaining of section 3 will detail all above steps in turn.

3.1 Initialization

In the initialization step, a population of μ parent individuals is randomly generated. The hybrid method is based on the generalized evolution strategy (GES) developed in [2] for mixed discrete-continuous variables and in which each individual is represented by four vectors (x_C, x_D, σ, p). σ and p are strategy parameters which control the application of mutation to respectively the continuous and discrete variables. In our hybrid method, the strategy parameters σ have not been used as deterministic methods control the creation of the continuous variables. Each individual is thus represented by three vectors (x_C, x_D, p). $p \in [0,1]^{nd}$ contains the mutation probabilities.

All the Lagrange multipliers and penalty parameters are set to 0 and r_0 respectively. r_0 is a value defined by the user.

3.2 Evaluation

During the last few years, several methods have been proposed for handling nonlinear constraints by evolutionary algorithms and for the evaluation of the individuals (see [7,6]). The most popular one is the method of static penalties because it is the simplest technique to implement. In this paper, we calculate the augmented Lagrangian function in order to evaluate each individual of the population. The advantage by using dual methods is that the property of global convergence is not limited to convex problems. It is extended to a larger number of problems [8].

3.3 Determination of the Discrete Variables of a New Individual

From steps 3.2a to 3.2b, a new population P'' of χ individuals ($\chi > \mu$) is created. Each offspring individual is generated by operations of recombination and mutation on the discrete variables (3.2a) and by a step in the continuous space along a Lagrangian based quasi-Newton direction for the continuous variables (3.2b).

The recombination operator used for the discrete variables is a discrete recombination in the global form:

$$x_{Di}^s{}' = x_{Di}^{p1} \text{ or } x_{Di}^{pi}, \, i = 1, \ldots, n_D \tag{10}$$

$$p1 \in [1, \mu], pi \in [1, \mu]$$

$p1$ and pi are individuals chosen at random in the population of parents. During the recombination operation for the creation of the discrete variables ($i=1, \ldots, n_D$) of an individual s, the first randomly parent $p1$ is held fixed for each discrete variable of the individual but the second parent pi is sampled anew for each discrete variable. This mechanism of recombination corresponds to the global form. For the probability of mutation, a global intermediate recombination is used ($p_i^s{}' = p_i^{p1} + 0.5(p_i^{pi} - p_i^{p1})$).

The mechanism of recombination allows a mixing of parental information and passes the information to the descendants.

The mutation introduces innovation. We have used the mutation mechanism proposed in [2] for the discrete variables and the mutation probabilities. It operates by first mutating the strategy parameters with a multiplicative, logarithmic, normally-distributed process. The mutated mutation probabilities($p_i^{s''}$) are then used to determine whether the discrete variable should be modified (x_{Di}^{s} "after mutation):

$$p_i^{s''} = \left(1 + \frac{1 - p_i^{s'}}{p_i^{s'}} \exp\left(-N_i(0,1)\left(\sqrt{2\sqrt{n_D}}\right)^{-1}\right)\right)^{-1} \tag{11}$$

3.4 Determination of the Continuous Variables of a New Individual

In the generalized evolution strategy developed in [2], the continuous variables of each offspring are created by recombination and mutation, like the discrete variables. In our hybrid algorithm, we determine the continuous variables of each offspring with deterministic methods, instead of creating them by recombination and mutation.

The details of the procedure A used for the creation of the continuous variables of the offspring individual s is summarized below.

x^0 denotes the starting point of the procedure: $x^0 = \{ x_C^{p1}, x_D^{s''} \}$.

$p1$ is the first randomly chosen parent used in the recombination mechanism for the creation of the discrete variables of the individual s. The parent $p1$ was held fixed during the creation of all the discrete variables of the offspring s (see §3.3).

(A.1) Determine the gradient of the augmented Lagrangian function with respect to the continuous variables at the point x^0.

(A.2) Calculate the Lagrange quasi-Newton direction

$$d = -H^{p1} \nabla L(x^0), d \in \mathbb{R}^{N_C}$$

H^{p1} is an approximation of the Hessian matrix of the augmented Lagrangian function, associated to the parent $p1$.

(A.3) Determine the step α by a line search in the direction d

(A.4) Calculate the continuous variables of the offspring s:

$$x_C^{s''} = x_C^{p1} + \alpha d$$

(A.5) Update the Hessian matrix of the augmented Lagrangian function, associated to the offspring individual s ($H^{s''}$) by the BFGS formula.

During the creation of the continuous variables of each offspring s, the discrete variables of this offspring determined by recombination and mutation ($x_D^{s''}$) are constant. The gradient of the augmented Lagrangian function with respect to the continuous variables is calculated at the point x^0 in order to define the search direction. A Lagrangian based quasi-Newton direction is used to define the search direction. The continuous variables of the offspring individual s are determined by the continuous variables of the parent individual $p1$ plus a step α along the search direction d. Several one-dimensional search methods exist to calculate the step α [8,5]. When the functions of the problem are monotonic, we have chosen to calculate this step by monotonicity analysis. The advantage of this method is that it is not an iterative procedure. The number of evaluations is therefore very low. When the functions of the problem are not monotonic, we determine the step with a golden section search method followed by a method based on quadratic interpolation.

In the procedure A, it is necessary to update the approximation of the Hessian of the augmented Lagrangian function, which will be associated to the offspring individual s. The BFGS formula [3] is employed by using $x^k = \{ x_C^{p1}, x_D^{p1})\}$ and $x^{k+1} = \{ x_C^s{}'', x_D^s{}'' \}$ for the update.

3.5 Selection

The selection drives the process toward better regions of the search space. A $(\mu + \chi)$ strategy has been used in step 3.2d where the μ best individuals from the union of the μ parents and χ offspring individuals are selected to be parents in the next generation. In order to keep some diversity in the population, we do not select more than two identical individuals.

3.6 Update of the Parameters

In step 3.3, the Lagrange multipliers are updated using the constraint functions evaluated at the best solution, corresponding to the best individual of the population. The penalty parameter is increased in step 3.4 by a constant rate defined by the user until a prespecified maximum value r_{max} is reached.

3.7 Termination Conditions

The step 3.2 allows a minimization of the Augmented Lagrangian function with respect to the primal variables (x). The termination condition of this step is defined by the verification of one of the following criteria:
• Maximum number of generation is reached.
• No improvement of the augmented Lagrangian function of the best individual during a fixed number of generations.

The step 3 allows a maximization of the dual function associated to the problem (see [8]) with respect of the dual variables (λ, r). The termination condition of this step corresponds to the verification of one the following criteria:
• Maximum number of generation is reached.
• The difference between the dual variables at the iteration (t) and the dual variables at the iteration $(t-1)$ with respect to the constraints with continuous variables is

small: ($\left| \lambda_j^t - \lambda_j^{t-1} \right| \leq \varepsilon \lambda_j^t$), and there is no difference between the discrete variables of the best feasible individual obtained with the dual variables λ_j^t and those of the best feasible individual obtained with the dual variables λ_j^{t-1}.

4 Application

For the application, we have chosen two complex mechanical design problems with mixed variables : the problem of a ball bearing pivot link (prob1) and the problem of a coupling with bolted rims (prob2).

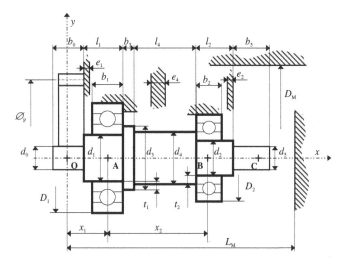

Fig. 1. Ball bearing pivot link

The aim of the first problem (see figure 1) is to find the lengths x_1, x_2 and the two ball bearings R_1 and R_2 in order to minimize the weight of the assembly composed of a shaft and two ball bearings. These ball bearings are chosen from a standardized table of prefabricated sizes. The list of geometrical conditions, stress conditions on the shaft, conditions on the bearing lifespan, maximum deformation of the shaft in bending, has allowed to formulate the optimization problem as a problem which finally contains 11 nonlinear inequality constraints, 2 continuous (x_1, x_2), 2 integer (R_1, R_2) variables and 14 discrete parameters (C_i, d_i, D_i, b_i, da_i, m_i, D_{wi}), $i=1..2$. R_1 and R_2 represent the choice of the two ball bearings. In order to solve this problem with the hybrid method, we numbered the ball bearings, having a diameter from 30 to 50, from 1 to 32, in the same order as the standardized table. (C_i, d_i, D_i, b_i, da_i, m_i, D_{wi}), $i=1, 2$ are respectively the parameters of the 2 ball bearings R_1 and R_2. They depend on the choice of the ball bearings. m_1 and m_2 represent the mass of the two ball bearings. The formulation of the optimization problem is presented in annex. The problem 2 contains

1 discrete variable (d), one integer variables (N), 2 continuous variables (R_B, M), 5 discrete bolt parameters and 3 nonlinear inequality constraints. Its equations are presented in [5].

The hybrid method, presented in this article is compared with an evolution strategy (ES). ES is used with the representation of the individuals proposed in [2] and the same recombination (for the discrete variables), mutation (for the discrete variables) and selection operators than those of the hybrid method. The choice of the recombination operators for the continuous variables and standard deviations has been realized by testing several recombination operators of the literature. Finally, a global extended generalized recombination ($x_i^{p1} + \alpha(x_i^{pi} - x_i^{p1})$), $\alpha \in$ [-0.5, 1.5)) has been employed for both the continuous variables and the standard deviations of ES.

The one-dimensional search method used to calculate the step with the hybrid method is the monotonicity analysis as the objective and constraint functions of each problem are monotonic with respect to the continuous variables in the search space of these problems. For the application, we have chosen a population of 7 parents and 50 offspring individuals. 15 runs are realized with each method in order to obtain statistically significant results. R_0 and γ have been chosen experimentally for each problem.

Finally, the hybrid method and the evolution strategy, applied to each design problem, have given the same best solution:

Prob1:   (R_1=14, R_2=5, x_1=35.5, x_2=68.84) $f^* = 728213.81$
Prob2:   (d=16, N=8, R_B=71.98, M=40) $f^* = 3.8796$

These results correspond to the best individual of the final population of the best run of each method. Thank to an enumerative method of all possible solution points of the problem, we have verified that these solutions are the global optimum of each mechanical problem.

Table 1. Results

		Aver. error (%)	total nb eval.	nb eval. optimum 10%
Prob1	Hybrid method	0.66	18805	367
	ES	2.35	37605	857
Prob2	Hybrid method	0.68	19305	340
	ES	2.58	37805	700

Table 1 gives, for each method and each problem, the average error of 15 runs for the final best objective function value with regard to the global optimum, the total number of evaluations of the objective function for 15 runs and the average number of evaluations of the objective function, which was necessary to obtain the global optimum with an error of 10%.

As we can see in table 1, we have obtained with the hybrid method results of better average reliability than those of ES alone, and with a smaller number of evaluations. The average number of evaluations needed to obtain a given reliability (the global optimum with a 10% error) is twice smaller with the hybrid method than with the evolution strategy.

The hybrid method has allowed to reduce significantly the computational cost and the average error for 15 runs is smaller. In this method, the evolution of the continuous variables is realized with adapted deterministic methods. Therefore, there is less exploration of the continuous search space than with ES, and the convergence is quicker. This hybrid method is particularly interesting for mechanical design problems which generally do not have local optima (ex: prob1 and prob2) and which have a small search space. The hybrid method allows to accelerate the convergence significantly in comparison with ES.

In order to test the performances of this hybrid method, we have compared this strategy with a deterministic method, based on the coupling of an enumeration of all the values of the discrete variables and the augmented Lagrangian method for the optimization of the continuous variables. About 100 evaluations of the objective function have been necessary to solve each mechanical design problem which contains 2 continuous variables, with the augmented Lagrangian method. As we have $32 \times 32 = 1024$ possible values of the discrete variables for the problem 1 and $8 \times 127 = 1016$ possible discrete values for the problem 2, a total number of about 102400 evaluations for the problem 1 and 101600 evaluations for the problem 2 is necessary to solve each mechanical design problem with the deterministic method. The computational cost of the hybrid method that we have proposed (18805 evaluations for the problem 1 and 19305 evaluations for the problem 2) is then smaller.

5 Conclusion

In this paper, we have proposed a hybrid method, based on the coupling of the evolution strategy with deterministic methods. Our aim was to use in parallel the advantages of evolutionary algorithms for the discrete variables and those of deterministic methods for the continuous variables. In the proposed hybrid method, the creation of a new individual is realized at each generation by a step in the discrete search space by mean of recombination and mutation, and a step in the continuous search space by mean of a move in a Lagrangian based quasi-Newton direction.

The hybrid method, applied to two strongly constrained mechanical design problems with mixed variables, has given in association with the monotonicity analysis interesting results. Compared with ES alone, the number of evaluations was reduced significantly and the reliability was improved.

It is now necessary to test the hybrid method on a large number of problems.

References

1. Bäck, T.: Evolutionary algorithms in theory and practice. Oxford University Press, New York (1996)
2. Bäck, T., Schütz, M.: Evolution strategies for mixed-integer optimization of optical multilayer systems. Proceedings of the Fourth Annual Conference on Evolutionary Programming, J.R. McDonnell, R.G. Reynolds, and D.B. Fogel, editors, MIT Press, Cambridge, MA, (1995) 33-51
3. Fletcher, R.: A new approach to variable metric algorithms. Computer Journal, vol.13 **3** (1970) 317-322
4. L. Giraud, P. Lafon: Optimization of mechanical design problems with genetic algorithms. Proceeding of the 2nd International Conférence IDMME'98, Compiègne, France, Mai, (1998) 98-90.
5. Lafon, P.: Conception optimale de systèmes mécaniques: Optimisation en variables mixtes. Thèse 3ème cycle, n° d'ordre 273, Institut National des Sciences Appliquées de Toulouse (1994)
6. Michalewicz, Z., Schoenauer, M.: Evolutionary algorithms for constrained parameter optimization problems. Evolutionary Computation **4** (1996) 1-32
7. Michalewicz, Z.: Genetic Algorithms + Data Structures = Evolution Programs. Springer, Berlin (1996)
8. Minoux, M.: Programmation Mathématique: Théorie et Algorithms. Tome1 et 2. Edition DUNOD (1983)
9. L. Moreau-Giraud, P. Lafon: Evolution strategies for optimal design of mechanical systems. Third World Congress of Structural and Multidisciplinary Optimization (WCSMO-3), Buffalo, Etats-Unis, Mai, (1999)
10. Myung, H., Kim, J.-H., Fogel, D.B.: Preliminary investigations into a two-stage method of evolutionary optimization on constrained problems. Proceedings of the Fourth Annual Conference on Evolutionary programming. J.R. Mac Donnel, R.G. Reynolds and D.B. Fogel (Eds), (1995) 449-463
11. Rockafellar, R.T.: Augmented lagrange multiplier functions and duality in nonconvex programmings. SIAM Journal Control, **12** (1974) 268-285
12. Schwefel, H.P.: Numerical optimization of computer models. Chichester: Wiley (1981)
13. Vasconscelos, J.A., Saldanha, R.R., Krähenbühl, L., Nicolas, A.: Genetic algorithm coupled with a deterministic method for optimization in electromagnetics. IEEE Transaction on magnetics, vol. 33 **2** (1997) 1860-1863
14. Wilde, D.J.: Monotonicity and dominance in optimal hydrolic cylinder design. ASME Journal of Engineering Optimzation, **97** (1975) 1390-1394
15. Zhou, J., Mayn, R.W.: Monotonicity analysis and the reduced gradient method in constrained optimization. ASME Journal of Mechanisms, Transmissions, and Automation in Design, **113** (1984) 90-94

Annex: Optimization Problem for the Ball Bearing Pivot Link(Prob1)

$$f(R_1,R_2,x_1,x_2) = m_1 + m_2 + \rho\frac{\pi}{4}\left[0.5\left(d_1^2(b_1-b_0)-(da_2)^2(b_1+b_2)\right)\right] \tag{12}$$

$$+ \rho\frac{\pi}{4}\left[(da_1)^2 b_3 + d_2^2 l_2 + x_1 d_1^2 + d_2^2(x_2-b_3)\right]$$

$$g_1(R_1,R_2,x_1,x_2) = 0.5(b_0+b_1) - x_1 + e_1 \le 0 \tag{13}$$

$$g_2(R_1,R_2,x_1,x_2) = 0.5(b_1+b_2) - x_2 + e_4 + b_3 \le 0 \tag{14}$$

$$g_3(R_1,R_2,x_1,x_2) = x_2 + x_1 + 0.5b_2 - L_M + e_2 + b_5 \le 0 \tag{15}$$

$$g_4(R_1,R_2,x_1,x_2) = \left(\frac{32K_{tff1}F}{pK_eK_sS_D}x_1 + \frac{16\sqrt{3}K_{tt1}C}{p(2R_m-s_D)}\right) - \frac{d_1^3}{a_F} \le 0 \tag{16}$$

$$g_5(R_1,R_2,x_1,x_2) = (60wL_V)^{1/3} F(1+x_1/x_2) - C_1 \le 0 \tag{17}$$

$$g_6(R_1,R_2,x_1,x_2) = (60wL_V)^{1/3} F(x_1/x_2) - C_2 \le 0 \tag{18}$$

$$g_7(R_1,R_2,x_1,x_2) = \left(\frac{F}{21490}\right)^{2/3}\left[\frac{(1+x_1/x_2)^{5/3}}{\left(Z_1\sqrt{D_{w1}}\right)^{2/3}} + \frac{(x_1/x_2)^{5/3}}{\left(Z_2\sqrt{D_{w2}}\right)^{2/3}}\right] \tag{19}$$

$$+ \frac{64F}{3pE}x_1^2\left(\frac{x_2}{d_{a2}^4} + \frac{x_1}{d_1^4}\right) - d_{max} \le 0$$

$$g_8(R_1,R_2,x_1,x_2) = D_2 - D_1 \le 0 \quad \textbf{(20)} \qquad g_9(R_1,R_2,x_1,x_2) = d_{min1} - d_1 \le 0 \tag{21}$$

$$g_{10}(R_1,R_2,x_1,x_2) = d_{min2} - d_2 \le 0 \quad \textbf{(22)} \qquad g_{11}(R_1,R_2,x_1,x_2) = D_1 - D_M \le 0 \tag{23}$$

$$\text{With}: d_{min1} = Max\left\{d_0,\left(\frac{16\sqrt{3}K_{tt1}Ca_S}{pR_e}\right)^{1/3}\right\}, d_{min2} = Max\left\{d_5,\left(\frac{16\sqrt{3}K_{tt2}Ca_S}{pR_e}\right)^{1/3}\right\} \tag{24}$$

b_0	30 mm	d_0, d_s	28 mm	ω	970 t/min
b_3	6 mm	ρ	7800 kg/m^3	K_e	0.83
e_1	8 mm	σ_D	425 MPa	K_s	0.96
e_4	7 mm	R_m	917 MPa	α_S, α_F	1.5
e_2	1 mm	R_e	845 MPa	δ_{max}	0.1 mm
D_M	100 mm	C	233.31 N.m	φ	20°
L_M	220 mm	F	6207.29 N	ϕ_p	80 mm
b_s	42 mm	L_V	1800 h		

Fig. 2. Data of the problem

Lamarckian vs Darwinian Evolution for the Adaptation to Acoustical Environment Change

Anne Spalanzani

CLIPS/IMAG Laboratory
BP 53
38041 Grenoble cedex 9
FRANCE
Anne.Spalanzani@imag.fr

Abstract. The adaptation to the changes of environment is crucial to improve automatic speech recognition systems' robustness in various conditions of use. We investigate the adaptation of such systems using evolutionary algorithms. Our systems are based on neural networks. Their adaptation abilities rely on their capacity to learn and to evolve. Within the framework of this work, we study both main methods concerning hybridization of training and evolution, namely the Lamarckian and Darwinian evolution. We show that the knowledge inheritance of a generation to another is much faster and more powerful for the adaptation to a set of acoustic environments changes.

1 Introduction

Speech recognition systems are faced to the problem of the conditions of use's variations. In particular, the acoustic environment changes regularly. The room's characteristics, its manner of reverberating sounds, the background noise caused by machines or people, etc. are all sources of variability [6]. The environment change adaptation is a problem for which Evolutionary Algorithms (EA) are well known to be adapted to. In order to maintain a proper control over this adaptation, two major problems have to be studied: the maintenance of diversity and the choice of the inheritance knowledge method. Within the framework of this article, we present works carried out to maintain sufficient diversity in the population and enable a good adaptation to the environment changes. We then present both methods characterizing the concept of heritage: evolution suggested by Lamarck and that proposed by Darwin.

We study these two methods in the context of the speech recognition. Initially (section 2), we present different ways of keeping a sufficient diversity enabling a good adaptation, in the second time (section 3), we present Lamarckian and Darwinian principles as well as the Baldwin effect. After having described the context of our simulations (section 4), we present our model in section 5 evolving a population of speech recognition systems. Results are presented in section 6.

C. Fonlupt et al. (Eds.): AE'99, LNCS 1829, pp. 136-144, 2000.

2   Adaptation to Environment Changes

The problem of adapting populations to changing environment is an interesting problem which gave place to a certain number of works.

For example, Nolfi and Parisi [11] investigated a robot adaptation in a changing environment controlled by a population of neural networks.

Cobb and Grefenstette [4] studied the evolution of a population tracking the optimum of complex functions (such as combination of sinusoids and gaussians) and the capacity of genetic algorithms to adapt, for example, to the translation of such functions. Keeping diversity in the population seems to be the key for a good adaptation to environments changing quickly. Cobb and Grefenstette [4] proposed a comparison of various strategies' performances (high mutation rate (10%), triggered hypermutation and random immigrants). The high mutation rate generates a significant diversity, the triggered hypermutation varies the mutation rate according to the way the environment changes, its rate is weak when the changes are weak, high during abrupt changes, the random immigrants introduces randomness into a percentage of its population which generates diversity. It results from this work that each of these methods has advantages and disadvantages depending on the way the environment changes.

Methods based on the thermodynamic principles [9] can be found in the literature also. In addition, evolution strategy seems to be well fitted to adapt to the changes of environment [2]. Indeed, the integration of evolution's parameters in the genotype enable the adaptation of the mutation rates when it is necessary. It is not very far from the idea of triggered hypermutation suggested by Cobb and Grefenstette [4].

3   Confrontation between Darwinism and Lamarckianism

A certain number of works studied the influence of the heritage on the evolution of the populations [15]. At the beginning of the century, two schools were confronted: The Darwinism is based on the idea that only the predisposition of the individuals to learn are transmitted to the children. Knowledge acquired by the parents is not transmitted then. Lamarckianism proposes that knowledge obtained by parents is directly transmitted to children. Hence, that parents' weights resulting from the training phase are transmitted to the following generation. In the context of this debate, Baldwin proposed a kind of intermediary [16]. He suggested the existence of a strong interaction between learning and evolution. The principal objective for the individuals is to integrate, by means of a neural network training, the information provided by the environment. The individuals having the best fitness, are selected for the reproduction. So they transmit to their descent their capacity to learn. If it is considered that an individual is all the more ready to integrate knowledge that the innate configuration of its neural network' weights is closed to that after training, we can consider then that knowledge is assimilated in its genes. Thus, the training time decreases. Once the individuals are able to acquire these concepts correctly, we can consider that those are comparable in the genotype.

Nolfi et al. [10] show that learning guides evolution (individuals having the best learning performances reproduce more often than the others), but also that evolution guides learning (the selected individuals have greater capacities to learn and these capacities are improved during generations). Thus, instinctive knowledge of the individuals is transmitted and improved during the evolution whereas, and it is the difference with the theory of Lamarck, the genotype is not affected directly by the training.

Whitley et al. [17] affirm that under all the test conditions they explored, the Lamarckian evolution is much faster than that of Darwin and results are often better. Concerning the problem in which we are interested in, the adaptation to the changes of environment, Sasaki and Tokoro [12] affirm that the Darwinism is more adapted than Lamarckianism, whereas for a static environment, the opposite is noted.

Mayley [8] proposes to penalise individuals having a long training phase. He affirms also that knowledge assimilation can be done only if there is a neighbourhood correlation, i.e. a correlation between the distance from two genotypes and that of their associated phenotype. In this article, we propose to compare both methods for our problem of adaptation to acoustic environment changes.

4 Background

The context of our work is speech recognition and more particularly the robustness of automatic speech recognition systems (ASRSs). An ASRS is robust if it is able to keep a good recognition rate even if the quality of the signal is degraded, or if the acoustic, articulatory characteristics or phonetics of the signal are different between the training phase of the system and the testing one. There are various sources of signal degradation but they can be classified in 3 kinds: speaker's variations (characteristic of the vocal tract for example), communication channels' variations (micro, telephone line, etc.) and acoustic environment's variations (reverberation, background noise, etc).

There are many methods to compensate the drop of performance of ASRSs caused by the mismatch between training and testing conditions, but very little are used in real ASRSs. In fact, most of these methods are still at the stage of experimental search and do not provide enough convincing results to be integrated. The most common method is certainly the increase of the database's size. As the computing power and the size of memory increase, it becomes possible to provide a good quantity of information for the training phase in order to have a powerful system in many conditions of use. However, in spite of the increase of their size, training databases are only small samples of the whole possible signal variabilities. It is not possible to forecast all the testing conditions are it is necessary to explore new ways of search for a better comprehension of the problems involved in speech recognition [3]. We chose to study the possible contributions of the evolutionary algorithms (EA).

In order to study the possibility of incorporating EA in the field of ASRSs' robustness, we propose to remain at the acoustic level of the speech processing and more particularly at the level of the ASRSs' adaptation.

In this context, ASRS's robustness can be approached by two ways: dealing with structure or dealing with stimuli.

The first approach consists in adapting corrupted testing data so that they become close to training data. The speech recognition system is then able to recognize them. In this study, the neural network's weights do not evolve anymore, data do only.

The second approach consists in adapting the speech recognition system itself. Through the combination of EA and Backpropagation, a kind of relearning is operated to adapt the system to an environment different from the one in which it has been trained. We studied the capacities of the system to adapt to acoustic conditions' changes using a local approach (by retro-propagation of the gradient) and a global one (by EA) in order to find an optimal ASRS. Results, presented in [13], show that evolution by EA improves ASRS population's results. The evolution used in these experiments was Lamarckian only. However, as the performances of Lamarckian and Darwinian evolution seem to depend on the studied application, a comparison of both type of heritage is necessary.

The goal of this experiment is to adapt speech recognition system to acoustic environment changes [13]. We present the model in the following section.

5   Experimental Design

The model presented below illustrates the method we use. Signals coming from the environment are analyzed by an acoustic analyzer PLP (Perceptual Predictive Analysis) [5]. The resulting acoustic coefficient vectors represent a database A. 80% of these data are for the training phase, 20% for the test. Individuals aim to learn and recognize these data, the best will be selected for the creation of the following generation.

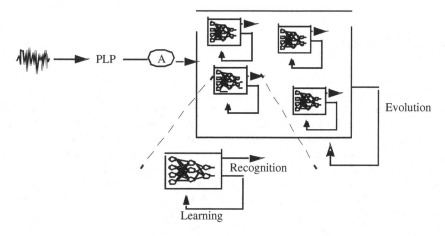

Fig. 1. Model for ASRSs' evolution.

A population of ASRSs is created randomly [7]. These ASRSs are connexionnist models based on feed-forward neural networks. The environments differ according to several criteria: the choice of the sounds (different vowels, different speakers), of the noises (alarm, radio, champagne, etc.), of the reverberation time (from 0 to 1 second) or of the signal to noise ratio (from -20dB to 20dB). The changes of environment are such as ASRSs are not adapted anymore to the new environment. A cycle of relearning by retro-propagation and the adaptation by EA is carried out.

```
1. Init a population of ASRSs.
2. If duration of simulation not elapsed change the environment
   else goto 6.
3. Train ASRSs.
4. Evaluate, Select and Reproduce ASRSs.
5. If duration of adaptation elapsed then goto 2 else goto 3.
6. End
```

Fig. 2. Evolutionary Algorithms for ASRSs' adaptation to new acoustic environments.

The model of virtual environment simulates the behaviour of sounds in a closed room. Thanks to the model of Allen on the propagation of the sound in small rooms [1], it is possible to determine the sound perceived by a microphone put somewhere in the room. The reverberation time, measuring the signal's time of extinction, is controlled by varying the walls' reflection coefficients. The higher it is and the more the signal is difficult to recognise.

6   Experimentation

The goal of these experiments is to determine the most effective method for our problem of adaptation to the environment. Since the opinions are divided concerning the method to use, we propose to test the performances of populations in term of quality of the individuals and in term of effectiveness. Within the framework of our experiments, this consists in studying the recognition rate of our population as well as the number of iterations necessary for each individual to optimise its training (i.e. the network converged). The objective of the individuals is to recognise a set of vowels analysed by an acoustic analyser based on the model of the human ear [5]. The vector resulting from the acoustic analysis represents the input of the networks having 7 input units, 6 hidden units and 10 output units. They are able to learn thanks to training algorithm based on the gradient retro-propagation. The evolution is carried out according to Genetic Algorithms with a crossing rate at 80% and a mutation one at 10%. Thus, with this operators' setting, we hopes to keep a diversity sufficient for a good adaptation.

We propose to compare the most common methods of evolution: the Lamarckian evolution and the Darwinian evolution. The population consists of 20 individuals put in an changing acoustic environment. For each environment, it has 100 generations to adapt. Figures 3 shows the performances of the populations evolving by Darwinian and Lamarckian methods of evolution (average on 9 simulations). Each curve shows the performances of the two extremes of the population (best and worst individual) as well as the average of the population.

The acoustic environments are represented by a serie of 200 vowels, to which noise and reverberation were added. Thus, an environment is defined by a triplet (type of noise, reverberation time, signal to noise ratio). The intelligibility of the signal is inversely proportional to the reverberation time and proportional to the signal to noise ratio. This is why a signal with a strong reverberation and a weak signal to noise ratio (for example (FE 0,7 -6)) is more difficult to recognise than a signal like (RA 0,4 20).

6.1 Quality of the Results

We can notice that the performances are quite equivalent in average. The numerical results of the performances are presented in table 1. In average, the population evolving according to the method of Lamarck obtains 78% of recognition rate whereas that according to the method of Darwin obtains 76.6%. Concerning the best individual, less than 1% of improvement is noted since in average (for the 10 environments, that is to say 1000 generations), the best Darwinian individual obtains 80.1% whereas Lamarckian 80.9%. On the general shape of the curves, we can also notice that results provided by the Lamarckian evolution are more stable and seem to drop less easily than those of the Darwinian population.

6.2 Training Efficiency

Concerning now the efficiency of the populations in the training phase, figures 4 shows the number of iterations necessary for a good convergence of the individuals' networks. Although, at each change of environment, the number of iterations increases in a more significant way during the evolution of Lamarck, the decrease of this number is more significant and, in average, the number of iterations is weaker. It is interesting to note that this number decreases throughout the Darwinian evolution, which can means that there is knowledge assimilation, and this without the use of penalty as proposed Mayley [8].

In both kinds of evolution, the number of iterations decreases during generations. Once more, we can notice that the evolution of Lamarck is more effective than that of Darwin. As indicates it table 2, Darwinian individual needs in average 91 iterations to learn correctly, whereas a Lamarckian individual need only 68.2 iterations. We can notice the differences between the best individuals (54.8 against 33.1) and worse (144 against 113.6).

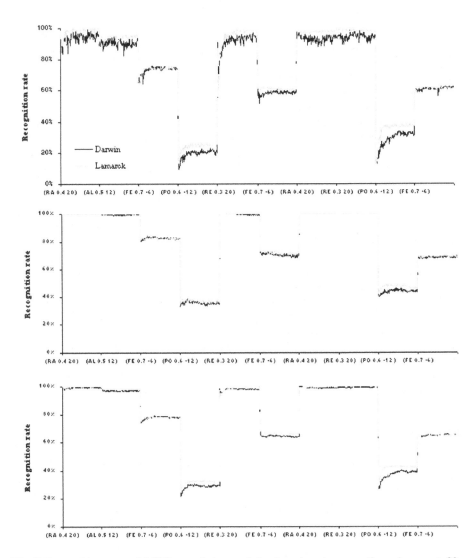

Fig. 3. Recognition rates of ASRSs population evolving in a changing acoustic environment. 20 individuals evolving by genetic algorithms and neural training. Darwinian and Lamarckian heritage are compared. Fig 3.a : Worst of the population. Fig 3.b : Best of the population. Fig 3.c : Average of the population.

Table 1. Comparison of the recognition rates.

Recognition rate	Worst	Best	Average
Lamarckian evolution	73.8 %	80.9 %	78 %
Darwinian evolution	70.5 %	80.1 %	76.6 %

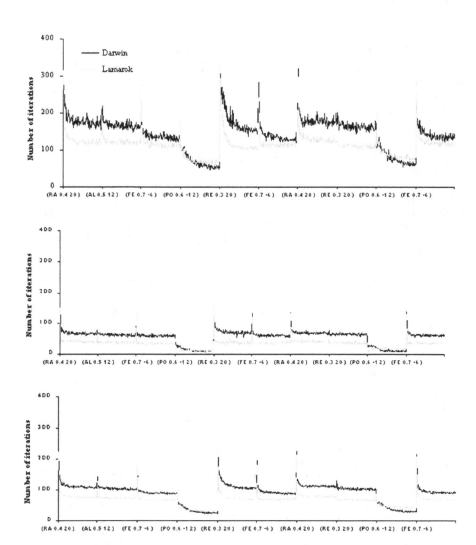

Fig. 4. Learning times of ASRSs population evolving in a changing acoustic environment. 20 individuals evolving by genetic algorithms and neural training. Darwinian and Lamarckian heritage are compared. Fig 4.a : Worst of the population. Fig 4.b : Best of the population. Fig 4.c : Average of the population.

Table 2. Comparison of the learning times.

Number of iterations	Worst	Best	Average
Lamarckian evolution	113.6	33.1	68.2
Darwinian evolution	144	54.8	91

7 Conclusion

We proposed in this article, to compare the methods of heritage suggested by Darwin and Lamarck for the evolution of a population in a changing acoustic environment. In term of speech recognition rate, both methods provide similar results. In term of learning speed, The Lamarckian evolution seems to be more interesting in the context of changing acoustic environments. These results join certain results presented by Whitley et al [17].

References

1. Allen J., B. & Berkley D. A.: Image Method for efficiently simulating small-room acoustics. JASA 65(4):943-950, 1979
2. Bäck T., Evolutionary Algorithms in Theory and Practice. Oxford University Press, New York, 1996
3. Bourlard H., Reconnaissance Automatique de la Parole : Modélisation ou Description ? XXIèmes Journées d'Etude sur la Parole(JEP'96), Avignon, France, pp.263-272, 1996
4. Cobb H.G. et Grefenstette J.J., Genetic Algorithms for Tracking Changing Environments. International Conference on Genetic Algorithms, Fifth International Conference on Genetic Algorithms (ICGA 93), 523-530, Morgan Kaufmann, July 1993
5. Hermansky, H.: Perceptual Linear Predictive (PLP) Analysis of Speech, Journal of Acoustic Society Am, 87(4) 1738-1752, 1990
6. Junqua, J. C. & Haton, H.: Robustness in Automatic Speech Recognition, Ed Kluwer Academic Publisher, 1996
7. Kabré, H. & Spalanzani A.: EVERA: A system for the Modeling and Simulation of Complex Systems. In Proceedings of the First International Workshop on Frontiers in Evolutionary Algorithms, FEA'97, 184-188. North Carolina, 1997
8. Mayley G., Landscapes, Learning Costs and Genetic Assimilation. Special Issue of Evolutionary Computation on the Baldwin Effect, vol. 4, n. 3, 1996
9. Mori N., Kita H. et Nishikawa Y., Adaptation to a Changing Environment by Means of the Thermodynamical Genetic Algorithm. 4th Conference on Parallel Problem Solving from Nature, Berlin, Germany, 1996
10. Nolfi S., Elman J.L., Parisi D., Learning and Evolution in Neural Networks. Technical Report 94-08, Department of Neural Systems and Artificial Life, Rome, Italy, 1994
11. Nolfi S. et Parisi D., Learning to Adapt to Changing Environments in Evolving Neural Networks. Technical Report 95-15, C.N.R. de Rome, Italy, 1996
12. Sasaki T. et Tokoro M., Adaptation toward Changing Environments : Why Darwinian in Nature ? Fourth European Conference on Artificial Life, 1997
13. Spalanzani A. & Kabré H., Evolution, Learning and Speech Recognition in Changing Acoustic Environments. 5th Conference on Parallel Problem Solving from Nature, Springer Verlag, pp. 663-671, Amsterdam, Netherland, 1998
14. Spalanzani A., Selouani S-A. & Kabré H., Evolutionary Algorithms for Optimizing Speech Data Projection. GECCO'99, Orlando, 1999
15. Turney P., Whitley D. et Anderson R., Evolution, Learning, and Instinct : 100 Years of the Baldwin Effect. Special Issue of Evolutionary Computation on the Baldwin Effect, vol. 4, n. 3, 1996
16. Turney P., The Baldwin Effect : A Bibliography. http://ai.iit.nrc.ca/baldwin/bibliography.html, 1996
17. Whitley D., Gordon S. et Mathias K., Lamarckian Evolution, The Baldwin Effect and Function Optimization. Parallel Problem Solving from Nature (PPSN III). pp. 6-15. Springer-Verlag, 1994

From Hough to Darwin: An Individual Evolutionary Strategy Applied to Artificial Vision

Jean Louchet

Ecole Nationale Supérieure de Techniques Avancées
32 boulevard Victor
75739 Paris cedex15, France
jean.louchet@ensta.fr
http://www.ensta.fr/~louchet

Abstract. This paper presents an individual evolutionary Strategy devised for fast image analysis applications. The example problem chosen is obstacle detection using a pair of cameras. The algorithm evolves a population of three-dimensional points ('flies') in the cameras fields of view, using a low complexity fitness function giving highest values to flies likely to be on the surfaces of 3-D obstacles. The algorithm uses classical sharing, mutation and crossover operators. The final result is a fraction of the population rather than a single individual. Some test results are presented and potential extensions to real-time image sequence processing, mobile objects tracking and mobile robotics are discussed.

1. Introduction

1.1. Segmentation and Scene Analysis

Mainstream computer vision and scene analysis techniques rely on the extraction of geometrical primitives from images ("image segmentation"), primarily based on pixel-level calculations. The goal of scene analysis may be viewed as an attempt to reconstruct a model of a three-dimensional scene, expressed in terms of geometric primitives and physical (e.g. photometric) attributes, generally using the results of segmentation (contours, regions, etc.) of one or several images, in order to build up the model.

One example is Stereovision, where the results of the segmentation of two or more images taken from different points of view, are compared and matched to exploit the (often small) geometrical differences between images and build a three-dimensional representation of the scene.

The results of scene analysis are often expressed in terms of a polyhedron-based description language, and used for instance in mobile robot planning applications for road tracking or obstacle avoidance. Impressive work and results have emerged from this approach, usually requiring heavy calculations.

C. Fonlupt et al. (Eds.): AE'99, LNCS 1829, pp. 145–161, 2000.

1.2.   Dual Space and the Hough Transform

An alternative to this approach to scene model reconstruction from images, has been given by Hough [7] and his many followers [1]. We can summarise the idea of the generalised Hough transform by considering a physical scene as a collection of objects, and defining the "dual space" as the space of the input parameters of a model able to represent any of these objects. The task of the Hough transform is to find which points (parameter vectors) in the dual space, would give the most likely explanation to features found in the given images. To this end, each image pixel votes for the subset of the dual space which consists of all the parameter vectors able to explain the pixel's features, or compatible with them. Once all pixels have given their vote, the subset of the dual space which contains the points with the highest number of votes is probably the parameter vector set we are looking after.

Unfortunately, in spite of several success stories [10], the generalised Hough transform suffers from its speed rapidly decreasing with the complexity of the patterns in the dual space. It becomes really unpractical with higher numbers of parameters, essentially because of the memory and time required to represent and value the full dual space.

1.3.   Artificial Evolution

If one considers the efficiency of the representation and exploration of the dual space, our view is that rather than calculating vote values everywhere, it will normally be cheaper to create and evolve a population of points in the dual space and let artificial evolution concentrate the population into the dual space points which would have obtained the highest votes in the Hough approach. This allows to only calculate the vote values on the points of the dual space where the population individuals are located.

The goal of this paper is to present one application of this approach to 3-D scene modelling. We chose the stereovision problem, which consists in building a 3-D model of the scene using the images taken by two cameras with known geometrical parameters. Robotics applications such as obstacle avoidance and path planning do not always require an exhaustive geometric description of the scene, which suffers from high computational costs in the "segmentation-based" methods, and where Hough methods are generally unpractical or useless due to the high dimension of the parameter space.

We will be using a fast individual evolution Strategy [3] to evolve a population of 3-D points in the space, in such a way that the population globally fits as accurately as possible the surfaces of the objects in the scene. "Individual" means that, unlike what happens with conventional evolutionary Strategies, the solution is not one "best" individual emerging from the population, but a large fraction of the population. Here, the output of the algorithm is a set of 3-D points and the dual space coincides with the physical space.

2.   Evolving Flies

2.1.   Geometry and Fitness Function

An individual (a "fly") in the population is defined as a 3-D point with coordinates (x, y, z). We are using two cameras, which we call the reference camera and the second camera[1]. The fly's projections coordinates are (x_R, y_R) in the image given by the reference camera and (x_S, y_S) for the second camera. The calibration parameters of the cameras are supposed to be known and allow to calculate x_R, y_R, x_S, y_S in function of x, y, z using the classical Projective Geometry formulas [6]:

$$
\begin{pmatrix} x_R \\ y_R \\ 1 \end{pmatrix} = F_R \begin{pmatrix} x \\ y \\ z \\ 1 \end{pmatrix} : \begin{pmatrix} x_S \\ y_S \\ 1 \end{pmatrix} = F_S \begin{pmatrix} x \\ y \\ z \\ 1 \end{pmatrix} \tag{1}
$$

where F_R and F_S are the projective $(4,3)$-matrices of the reference and second cameras: for example,

$$
F_L = \begin{bmatrix} r_{11} & r_{12} & r_{13} & t_1 \\ r_{21} & r_{22} & r_{23} & t_2 \\ r_{31} & r_{32} & r_{33} & t_3 \end{bmatrix} \tag{2}
$$

where r_{ij} are the elements of an orthogonal (rotation) matrix and t_i are translation terms.

In order to simplify calculations, we have chosen the reference camera's coordinate system as the general coordinate system. Thus, the z-axis is the reference camera's axis, and:

$$
F_R = \begin{bmatrix} 1 & 0 & 0 & 0 \\ 0 & 1 & 0 & 0 \\ 0 & 0 & 1 & 0 \end{bmatrix} \tag{3}
$$

The essential idea of the algorithm is the following. If the point is located on the surface of an object, then the corresponding pixels in the two images will very probably have the same grey levels[2]. Conversely, if the fly is not on the surface of an object, thanks to the non-uniformity of objects and illumination, the grey levels of its projections will have no reason to get the same grey level. The algorithm presented translates this property into a fitness function and evolve the flies' population from a random initial population. The population is initialised in order to occupy the space in the intersection of the cameras fields of view, from a given minimal distance to infinity. The fitness function evaluates the degree of similarity of the pixel neighbourhoods of the projections of the fly onto each image: this ensures highest fitness values for the individuals lying on the surface of an object.

[1] In the standard Stereovision case, they are the left and right cameras
[2] This is essentialy true with Lambertian (matt) surfaces where rediffusion of incident light is isotropic. Most usual non glossy surfaces slightly from the Lambertian model, but this may be at least partly taken into account in the fitness function (see below). Reflections on glossy surfaces may rise to virtual objects and wrong 3-D interpretation, independently oh the class of image processing algorithm being used.

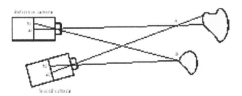

Fig. 1. : pixels b_1 and b_2 , projections of fly B, have identical grey levels. Pixels a_1 and a_2 , projections of fly A, do not necessarily have identical grey levels as they correspond to two different points on the surface of the object

As shown in Fig. 1, the fitness function, if only taking into account the dissimilarity between the two projections of a fly, would give unwanted high fitness values to flies located in front of a uniform object, even if not on the object's surface. In order to overcome this problem, the fitness function, which would otherwise be defined as a measurement of the similarity of immediate neighbourhoods of the corresponding pixels[3], has to include a normalizing term in the numerator:

$$fitness(indiv) = \frac{G}{\sum_{(i,j)\in neighborhood}(R(x_R + i, y_R + j) - S(x_S + i, y_S + j))^2} \qquad (4)$$

where:
- $R(x_R + i, y_R + j)$ is the grey level of the reference image at pixel $(x_R + i, y_R + j)$
- $S(x_S + i, y_S + j)$ is the grey level of the second image at pixel $(x_S + i, y_S + j)$
- N is a small neighbourhood, in order to measure the match quality over several pixels.

The normalizing numerator G gives a measurement of the mean local contrast, based on a gradient calculation, and can be precalculated. Our experiments showed that defining G as

$$G = \sqrt{\sum_{(i,j)\in neighborhood}(R(x_R + i, y_R + j) - R(x_{R,} y_R))^2} \qquad (5)$$

provides a good trade-off between giving high fitness values to non significant pixels and giving undue advantage to highly contrasted ones. In practice, it only needs to be calculated using the grey levels from one of the two images.

Additionally, the fitness function is slightly altered in order to reduce its sensitivity to lower spatial frequencies, through the application of a linear filter (subtracting a local mean) to the images.

Thus, most pixel-level calculations are contained in the fitness function. Lets us now examine the operators of the evolutionary resolution engine.

2.2. Artificial Evolution

The *initial population* is generated in the vision cone of the reference camera, truncated using an arbitrary clipping distance. An individual's *chromosome* is the triple (x, y, z) which contains the individual's coordinates in the coordinate system, O_Z being the camera axis. The statistical distribution of the individuals is chosen in order to obtain a uniform distribution of their projections in the reference image. In addition, we choose a uniform distribution of the values of z^{-1} such that the individuals stay beyond an arbitrary clipping line (minimum distance): this implies that the individuals' probability density is lower at high distances. The geometrical calibration parameters of the cameras are supposed to be known. This allows, for each individual (x, y, z), to calculate its image coordinates in each camera and calculate the corresponding fitness value.

---

[3] The denominator measures a square distance between the pixel configurations around the projections of the individual on the two images. Thus, highest fitnesses are obtained for individuals whose projections have similar b*ut significant* pixel surroundings.

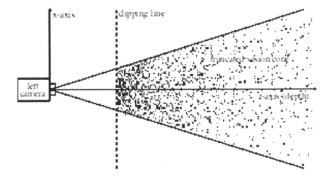

Fig. 2. the fly population is initialised inside the grey region of the 3-D space (truncated vision cone).

Selection uses a fast ranking process, based on the individuals' fitness values. This results into an approximate ranking, depending on the number of histogram steps chosen. In order to prevent the population from getting concentrated into a very small number of maxima, a 2-D *sharing process* allows to reduce the fitness values of individuals located in crowded areas. Thus, the presence of one individual A with coordinates (x, y, z) lowers the fitness values of all the individuals whose projection on the reference image is close enough to the projection of individual A.

The *mutation* operator is a quasi-Gaussian noise added to the individuals' chromosome parameters X, Y and Z, with a fixed standard deviation.

A *crossover* operator will be introduced in the next Section.

3.   Evolutionary Operators

3.1.   Sharing

The following synthetic image colour stereo pair[4] shows several discs at various distances, with a flat wooden wall in the background. We choose the left image as the "reference image" and the right image as the "second image".

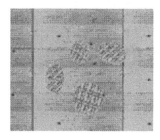
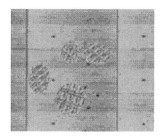

Fig. 3 : "Money image", left.   Fig. 4 : "Money image", right.

Figures 5 and 6 show the results of the 30% best individuals after 50 generations, without sharing.

Fig. 5 is a representation of the scene as it could be seen from above (vertical projection of the population), showing the axes x_l (horizontal) and $1/z$ (vertical) - the horizontal dotted line corresponds to an infinite distance $(1/z = 0)$. Fig. 6 is a range image obtained with the same population,where grey levels represent the values of z (darker pixels correspond to flies at shorter distances). We used a population of 5000 individuals, a mutation rate of 60%, and no crossover. The 1500 best individuals (30%) are displayed after 50 generations. The flies tend to concentrate into pixels with highest gradient values, even with higher numbers of generations.

[4] "Money" image pair, ©INRIA - Mirages project.

Fig. 5 : vertical projection (top view)   Fig. 6 : front view (darker is closer)

To correct this, we implemented a sharing operator [5] using a partition of the image into square regions and counting the number of flies whose projections are within each square. Each fly's fitness is then decremented proportionally to the population of the square. Figures 7 and 8 give the results with a square size ("sharing radius") of 2 pixels and a sharing coefficient of 0.2 ("medium sharing"), Figs. 9 and 10 with a sharing radius of 3 and a coefficient of 0.5 ("high sharing"). In Figs. 7 and 8, the wall in the background and the four discs positions are detected in a fairly acceptable way. However, noisy results in Figs. 9 and 10 are the consequence of excessive sharing.

Fig. 7 : top view, medium sharing   Fig. 8 : front view, medium sharing

Fig. 9 : top view, high sharing   Fig. 10 : front view, high sharing

3.2. Crossover

Many real-world images contain convex primitives as straight lines or planar surfaces. We translated this feature into a *barycentric crossover operator* which builds an offspring randomly located on the line segment between its parents: the offspring of two individuals with space coordinates (x_1, y_1, z_1) and (x_2, y_2, z_2) is the individual whose space coordinates (x_3, y_3, z_3) are defined by :

$$x_3 = \lambda x_1 + \mu x_2 ; y_3 = \lambda y_1 + \mu y_2 ; z_3 = \lambda z_1 + \mu z_2 \tag{6}$$

where the weights λ, μ are chosen using a uniform random law typically in the [0,1] interval[5] $(\lambda + \mu = 1)$.

The results on 100 generations with the same test images as above, with a population of 5000 individuals, show the effect of different crossover rates (Figs. 11 - 16).

[5] It is generally accepted that such a crossover operator has contractive properties which may be avoided by using a larger interval. However the idea in our application is that contrasts are often higher on objects' edges and therefore higher fitness values and higher individuals densities are likely to be obtained on objects' edges. The goal of the crossover operator is to fill in surfaces whose contours are easier to detect, rather than to extend them. It is therefore not always desirable to use coefficients allowing the centre of gravity to lie outside of the object's boundary.

Fig. 11: top view, no crossover   Fig. 13: medium crossover rate   Fig. 15: high
                                          crossover rate   crossover rate

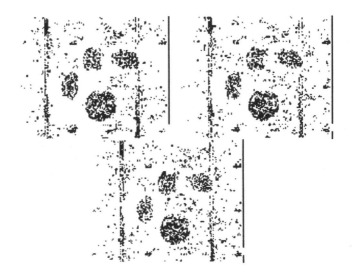

Fig. 12: front view, no crossover   Fig. 14: medium crossover rate   Fig. 16: high crossover rate

The following table gives the details of parameter values used in Figures 11 - 16.

Figures 11,12:	Figures 13,14:	Figures 15,16:
population 5000	population 5000	population 5000
100 generations	100 generations	100 generations
mutation 60%	mutation 40%	mutation 20%
no crossover	crossover 20%	crossover 40%
sharing radius 2	sharing radius 2	sharing radius 2
sharing coefficient 0.3	sharing coefficient 0.2	sharing coefficient 0.2
CPU 8.47 sec.	CPU 8.35 sec.	CPU 8.53sec.
298959 evaluations	298849 evaluations	299049 evaluations
40% displayed	40% displayed	40% displayed
mean fitness 2.891	mean fitness 2.949	mean fitness 2.903

In the vertical projection shown in Figures 11, 13, 15, the fuzzy aspect of the fourth disc on the right is not due to the algorithm but to the display mode used here : unlike the other ones it is not contained in a vertical plane.

A slightly better average fitness value is obtained here with 40% mutation and 20% crossover rates, but this depends on the test images chosen, and also on the population size as shown below. Higher crossover rates tend to fill in some spaces between objects.

The trade-off between mutation and crossover rates does not give very significant differences in the above example, but we will see below how it can make a very significant difference when using small populations.

CPU time shown is based on a 366MHz Linux i686 PC and may vary (typically ?10%). It includes the I/O operations. Without I/O and initialisation operations, one generation takes about 60 milliseconds, with a population of 5000.

When smaller populations are used e.g. in order to reduce computation time, the sharing coefficient must be increased to ensure a fair repartition of the population. However, particularly in this case, it appears that the introduction of crossover allows better exploration of the search space and more reliable object detection. The next examples (Figs. 17 - 19), using 1000 individuals, show how the crossover rate can affect object detection.

Fig. 17: small population, no crossover   Fig. 18: medium crossover   Fig. 19: high crossover

The parameter values used in Figs. 17 - 19 are given in the table below.

population 1000	population 1000	population 1000
100 generations	100 generations	100 generations
mutation 60%	mutation 40%	mutation 20%
no crossover	crossover 20%	crossover 40%
sharing radius 3	sharing radius 3	sharing radius 3
sharing coefficient 1.0	sharing coefficient 1.0	sharing coefficient 1.0
CPU 2.04 sec.	CPU 2.01 sec.	CPU 2.03 sec.
58831 evaluations	58451 evaluations	58704 evaluations
80% displayed	80% displayed	80% displayed

4.   Parameter Sensitivity and Convergence Results

We present results with the synthetic "Money" test images, as they correspond to a simple scene and allows easier readability of results. However, very similar results are obtained with natural images.

4.1.   Mutation vs. Crossover and Mean Fitness

Using the same parameter values as in Figures 11 - 16, we obtained the evolution of fitness values, averaged on the whole population, shown on Fig. 20.

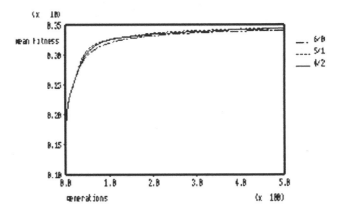

Fig. 20 : Average fitness values for a 5000 individual population in function of generations, using three different mutation/crossover combinations.

The best average fitness values are obtained with a 50% mutation rate and a 10% crossover rate. The bottom curve corresponds to a mutation-only evolution.

4.2.   Number of Generations

Results on the Money image after 10, 50, 100 and 1000 generations, using a 50% mutation rate and a 10% crossover rate are shown on Figs. 21 - 24.

Fig. 21 : 10 generations   Fig. 22 : 50 generations

Fig. 23 : 100 generations   Fig.24 :1000 generations

4.3. Results on Real Images

4.3.1. Standard Stereovision Images

The 760 ? 560 image pair on Figs. 25 and 26 has been taken using a single monochrome video camera with a sideways translation movement. The left image has been chosen as the reference image, the right image as the second image. We used similar genetic parameters (5000 individuals, 100 generations, 40% mutation, 20% crossover, sharing radius 2, sharing coefficient 0.3).

Fig. 25: left image   Fig. 26 : right image

Fig. 27 : top view (from 768 ? 576 pixel images)Fig. 28 : top view (from 384 ? 288 pixel images)

On Fig. 27 and 28, one can see the two sides of the cabinet, the beginning of the wall on the right and the front half-circle of the stool[6].

4.3.2. Axial Vision Images

The following image pair has been taken using the same video camera as above, but with a forward motion along its axis. The calibration matrix has been modified accordingly. The Focus Of Expansion (FOE) is at the image centre. We used the same genetic parameters as in Section 4.3.1.

Fig. 29: reference image   Fig. 30: second image

In these complex scenes the vertical projection of the population (Fig. 31) is more difficult to read, but the range image (Fig. 32) shows an adequate detection of the boxes on both sides, including the specular reflection of the box on the left. The distances of the wall and other objects in the background are estimated properly. The rails used as camera guides are not detected, as they are epipolar lines [6] converging to the FOE and give no useful information. Similarly, there is an empty area around the image centre, as the apparent velocities near the FOE are low and do not provide accurate information.

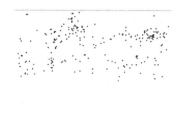

Fig. 31 : top view   Fig. 32 : range image (darker is closer)

5.   Towards Real-Time Evolution: Tracking Moving Objects

It is often considered that Genetic Algorithms and evolution Strategies are slow and therefore not well adapted to real-time applications. However :

- Speed is not the only issue with real-time applications. Real-time means the ability to exploit incoming data and react to them as fast as needed by the end user. Evolution Strategies are generally able to adaptation, i.e. to cope with modifications of the fitness function during the algorithm's execution [12], unlike most other optimisation methods.
- The processing speed of an evolutionary Strategy is strongly dependent on the computational complexity of the fitness function - which is fairly simple in our case.

We are now extending our algorithm to image sequences and stereo pair sequences. In order to be able to process objects' movements more efficiently we are using a larger chromosome, which includes the particle's speed vector. Mutation now becomes a quasi-Gaussian noise applied to speeds rather than coordinates: this allows the population to keep in memory the velocities of the objects to be tracked. The crossover operator is modified accordingly.

---

[6] It is to be noted that (as it is often the case in computer vision applications) there are no simple means of comparison with results obtained from conventional stereovision algorithms which use primitives as line segments or regions. However on a qualitative viewpoint, our results can be considered good compared to conventional methods.

6.   Conclusion

We described a fast evolutionary Strategy able to give a three-dimensional description of a scene from stereo images. Unlike conventional, computation-extensive approaches to stereovision, this progressive method does not require any image pre-processing or segmentation. The gradually increasing accuracy of results should be of interest in Robotics applications, to process image sequences containing moving objects.

While the Hough transform uses a vote technique in order to explore the parameter space, with our method the parameter space is explored by the evolving population, each individual testing some pixel properties through a model of image formation. There is no obvious general rule telling which is the most efficient approach, but it appears that even with the flies' example, with only three parameters, a complete Hough-style filling up of the 3-dimensional parameter space would already result in much higher complexity. The benefits of the evolutionary approach are:

- high speed processing[7] (partly due to the non-exhaustive search),
- progressive accumulation of knowledge about the scene, making it possible to use the results at any stage of the algorithm,
- real-time compliance, as the fitness function may be defined using external variable sensor-based data (e.g. in mobile Robotics applications).

We are currently introducing stereo camera motion into the algorithm in order to exploit information accumulated from the preceding image pairs and estimate the objects' and observer's speeds. Our plans are to implement this evolutionary scene analysis method into a low-cost mobile robot vision application [4], and design a motion planning system that uses directly the 3D particle-based representation of the scene as an input. Other applications are being undertaken in our team, in particular in positron emission tomography image reconstruction, introducing a model of photon diffusion into an otherwise similar individual evolution strategy [2].

7.   Bibliography

[1] Dana H. Ballard, Christopher M. Brown, *Computer Vision*, Prentice Hall, 1982.
[2] Lionel Castillon,. Correction de la diffusion Compton par algorithmes génétiques, internal report, ENSTA, June 1999.
[3] Pierre Collet, Evelyne Lutton, Frederic Raynal, Marc Schoenauer, *Individual GP: an Alternative Viewpoint for the Resolution of Complex Problems,* GECCO99, Orlando, Florida, July 1999.
[4] D. B. Gennery, Modelling the Environment of an Exploring Vehicle by means of Stereo Vision, PhD thesis, Stanford University, June 1980
[5] David E. Goldberg, Genetic Algorithms in Search, Optimization and Machine Learning, Addison Wesley, 1989.
[6] R. M. Haralick, *Using Perspective Transformations in Scene Analysis*, Computer Graphics and Image Processing 13, 1980, pp. 191-221.

[7] P. V. C. Hough, *Method and Means of Recognizing Complex Patterns*, U.S. Patent n°3, 069 654, 18 December 1962.

[8] Evelyne Lutton, Patrice Martinez, A Genetic Algorithm for the Detection of 3D Geometric Primitives in Images, 12th ICPR, Jerusalem, Israel, October 9-13, 1994 / INRIA technical report # 2210.

[9] David Marr, *Vision*, W.H Freeman and Co., San Francisco, 1982.

[10] John O'Rourke, *Motion Detection using Hough technique*, IEEE conference on Pattern Recognition and Image Processing, Dallas 1981, pp. 82-87.

[11] G. Roth and M. D. Levine, *Geometric Primitive Extraction using a Genetic Algorithm*, IEEE CVPR Conference, pp. 640-644, 1992.

[12] Ralf Salomon and Peter Eggenberger, *Adaptation on the Evolutionary Time Scale: a Working Hypothesis and Basic Experiments*, Third European Conference on Artificial Evolution, Nîmes, France, October 1997, Springer Lecture Notes on Computer Science no. 1363, pp. 251-262.

---

[7] processing speed does not depend directly on image size but on population size.

A New Genetic Algorithm for the Optimal Communication Spanning Tree Problem

Yu Li and Youcef Bouchebaba

LaRIA, Univ. de Picardie Jules Verne
33, Rue St. Leu, 80039 Amiens Cedex, France
fax: (33) 3 22 82 75 02
{yli@laria.u-picardie.fr, boucheba@laria.u-picardie.fr}

Abstract. This paper proposes a new genetic algorithm to solve the Optimal Communication Spanning Tree problem. The proposed algorithm works on a tree chromosome without intermediate encoding and decoding, and uses crossovers and mutations which manipulate directly trees, while a traditional genetic algorithm generally works on linear chromosomes. Usually, an initial population is constructed by the standard uniform sampling procedure. But, our algorithm employs a simple heuristic based on Prim's algorithm to randomly generate an initial population. Experimental results on known data sets show that our genetic algorithm is simple and efficient to get an optimal or near-optimal solution to the OCST problem.

1 Introduction

A key issue for designing a genetic algorithm (GA) is how to represent the solution space. A traditional GA (e.g. the Simple Genetic Algorithm of Goldberg (1989)) generally uses a fixed representation space, which consists of binary strings or vectors, called *linear chromosome*. Linear chromosome is natural for problems defined directly over binary strings or vectors. However, for problems in which possible solutions are more complicated objects, e.g. the Optimal Communication Spanning Tree (OCST) problem addressed in this paper, for which a solution is a tree, a linear chromosome may be unnatural or even ineffective.

Several researchers have felt the importance of incorporating domain knowledge into the representation of the solution space (Radcliffe & Surry 1995; Wolpert & Macready 1995). Only one practical approach, to our best knowledge, was developed by Sinclair (1998) for optical network optimization, modeling and design (NOMaD), in which network objects are themselves the structures undergoing adaptation and a variety of specific-problem operators have been incorporated within NOMaD.

Although linear chromosome seems to unnatural or even ineffective for incorporating domain knowledge into GAs, there is still a continued interest to follow this traditional framework of GAs in some ways. This is the case in the field of tree network design. Several tree encodings have been suggested to code a spanning tree, such as Predecessor (Berry *et al.* 1995) , Prüfer number (Julstrom

C. Fonlupt et al. (Eds.): AE'99, LNCS 1829, pp. 162–173, 2000.

1993), a tree encoding based on determinant factorisation (Abuali *et al.* 1995), and a more complicated chromosome encoding involving bias values for nodes and links (Palmer & Kershenbaum 1995). The advantages and disadvantages of these encodings have been discussed in detail in their papers.

Another important issue for designing a GA is the initialization procedure. In a traditional GA, an initial population is chosen uniformly at random from the whole search space. However, all GA practitioners have observed that a bad initialization procedure may, in the best case, modify the online performance (i.e. increase the time-to-solution), and, in the worst case, prevent the convergence towards the global optimum.

This important issue drew the attention of some researchers (Surry & Radcliffe 1996; Kallel & Schoenauer 1997). It is argued that the standard uniform sampling procedure could be improved by incorporating the domain knowledge into the initialization procedure.

In this paper, we propose a new GA to solve the OCST problem, in which the tree structure itself is taken as the chromosome without intermediate encoding and decoding. The crossovers and mutations directly manipulate trees, and an initial population is generated randomly by a simple heuristic based on Prim's algorithm rather than uniformly at random in the whole solution space. Experimental results on known sets show that our GA is simple and effective to get an optimal or near-optimal solution to the OCST problem.

Our experiences suggest that the GA could work directly on the representations which capture the structure of problems rather than on a traditional linear chromosome. We also feel that an initial population of "good quality" could greatly improve the performance of GAs.

The paper is organized as follows. A formal definition of the OCST problem is given in Section 2. Section 3 presents our tree chromosome, the crossover, and mutation operators. Section 4 describes the initialization procedure. We give an overview of our GA in Section 5. The numerical results are presented in the same section. We make some concluding remarks in Section 6.

2   Problem Definition

The problem of finding a spanning tree which minimizes the cost of transmitting a given set of communication requirements between n sites over the tree edges was introduced by Hu (1974) and referred to as OCST problem. It was also referred to as Minimum Communication Cost Spanning Tree problem (see Crescenzi *et al.* 1998). This problem arises directly in the design of tree communication networks. It was proved NP-complete by Johnson *et al.* (1978).

Formally, we are given a complete undirected graph $G = (V, E)$, the distance $d_{i,j}$ between each pair of nodes, and the communication requirement $r_{u,v}$ for each pair of nodes. We assume that the corresponding requirement matrix R is symmetric or uppertriangular, and that the corresponding distance matrix D is symmetric or uppertriangular, but it need not obey the triangle inequality. We wish to find a spanning tree of minimum communication cost.

The communication cost of a given spanning tree is defined as follows. For a pair of nodes u and v, there is a unique path in the spanning tree between u and v. The distance of the path is the sum of distances of edges in the path. The communication cost for the pair of nodes u and v is $r_{u,v}$ multiplied by the distance of the path. Summing over all $\binom{2}{n}$ pairs of nodes, we have the cost of spanning tree. The objective function is then

$$\min_{T \in \Im} \left[\sum_{u,v \in V} \left(r_{u,v} \sum_{(i,j) \in P_{u,v}(T)} d_{i,j} \right) \right]$$

where $\Im$ is the set of n^{n-2} labelled trees on n nodes (Gibbons 1985), and $P_{u,v}(T)$ is the unique path between u and v in the tree T.

For example, given a network of six nodes, the distances and the requirements between the six nodes are as follow

$$\mathcal{D} = \begin{pmatrix} 0 & 3 & 6 & 5 & 9 & 7 \\ 0 & 0 & 3 & 2 & 4 & 8 \\ 0 & 0 & 0 & 3 & 7 & 2 \\ 0 & 0 & 0 & 0 & 9 & 2 \\ 0 & 0 & 0 & 0 & 0 & 1 \\ 0 & 0 & 0 & 0 & 0 & 0 \end{pmatrix} \qquad \mathcal{R} = \begin{pmatrix} 0 & 5 & 13 & 12 & 8 & 9 \\ 0 & 0 & 7 & 4 & 2 & 6 \\ 0 & 0 & 0 & 3 & 10 & 15 \\ 0 & 0 & 0 & 0 & 11 & 7 \\ 0 & 0 & 0 & 0 & 0 & 12 \\ 0 & 0 & 0 & 0 & 0 & 0 \end{pmatrix}$$

The cost of the spanning tree in Figure 1, $cost(T)$, is

$$
\begin{aligned}
cost(T) &= r_{12}(d_{12}) + r_{13}(d_{12} + d_{24} + d_{46} + d_{63}) + r_{14}(d_{12} + d_{24}) + \ldots + r_{46}(d_{46}) + r_{56}(d_{56}) \\
&= 5(3) + 13(3 + 2 + 2 + 2) + 12(3 + 2) + \ldots + 7(2) + 12(1) \\
&= 534
\end{aligned}
$$

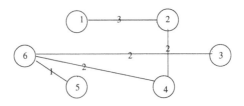

Fig. 1. A spanning tree of six nodes

Note that this problem is different from the familiar Minimum Spanning Tree (MST) which is computationally easy (algorithms of Kruskal, Prim).

Hu (1974) gave exact solutions for two restricted cases of this problem, and Peleg (1998) provided approximate solution to several other cases of the problem. However, for the general case of the problem, we only find two GAs (Palmer & Kershenbaum 1995; Berry *et al.* 1995) to obtain near-optimal solutions.

3 Tree Chromosome, Crossover and Mutation

We use tree itself as chromosome in our GA. An individual in the population is then a tree. This chromosome representation requires new crossovers and mutations to directly manipulate trees.

3.1 Crossover

By blending traditional uniform crossover with tree chromosome representation, we obtain two crossovers. One is based on path exchange called *path crossover*, and another on subtree exchange called *tree crossover*.

Path crossover firstly generates k random paths in two trees (parents) (k is a random number), then exchanges these paths between two trees by inserting a path of a tree into another tree and at the same time deletes cycles produced during the insertion.

The procedure of *path crossover* is described in Figure 2.

Procedure pathCrossover($parent1, parent2, child1, child2$)
Begin
```
    choose a random number k
    for i = 1 to k do
    begin
```
      $path1$ = randomPath($parent1$)
      $path2$ = randomPath($parent2$)
      insertPath($path1, child2$)
      insertPath($path2, child1$)
```
    end
```
End.

Fig. 2. Path crossover procedure

An example is given in Figure 3 to illustrate *path crossover*.

Tree crossover consists of exchanging subtrees between two trees which similarly works with *path crossover*.

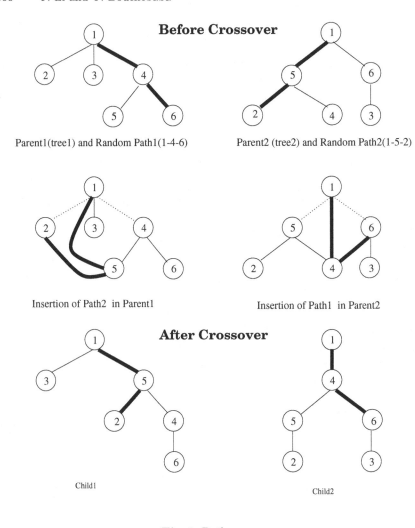

Fig. 3. Path crossover

3.2   Mutation

Mutations in GAs aim to create diversity that may not have been present in the initial population. We propose a mutation called *path mutation* which extracts random paths from the complete graph, and insert them in a child obtained after crossover. The procedure of *path mutation* is described in Figure 4, and an example of *path mutation* is shown in Figure 5.

tree mutation extracts random subtrees from the complete graph, and inserts them in a child obtained after crossover.

Procedure pathMutation(*child*)
Begin
    `choose a random number` k
    `for` $i = 1$ `to` k `do`
    `begin`
        path = `randomPath`(*completeGraph*)
        `insertPath`(*path*, *child*)
    `end`
End.

Fig. 4. Path mutation procedure

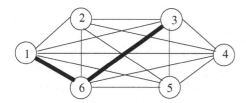

Complete Graph and Random path (1-6-3)

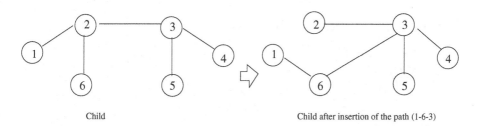

Child   Child after insertion of the path (1-6-3)

Fig. 5. Path mutation

3.3   Complexity of Crossovers and Mutations

The generation of a random path or a random subtree can be effected by a modified depth-first search. The insertion of a path or a subtree in a tree is in fact the insertion of the edges of the path or the subtree. We describe the procedure of inserting an edge in Figure 6.

Therefore, the main operations of crossovers and mutations consists in searching a random or determinist path. The complexity of crossovers or mutations is then proportional to the number of searching a path multiplied by the complexity of searching a path.

Procedure insertEdge(*tree, e*)
Begin
    if $e = (v1, v2)$ is **not** in *tree*
    then
        path = deterministPath(*tree*, *v1*, *v2*)
        delete an edge on *path*, which is **not** recently inserted
        add *e* to *tree*
    end
End.

Fig. 6. Edge insertion procedure

4   Initial Population

We propose a simple heuristic based on Prim's algorithm to randomly generate a spanning tree of a relatively small cost. We describe this heuristic in Figure 7.

Procedure generateSpanningTree()
Begin
    randomly choose a $u \in V$
     $T' \leftarrow \emptyset$
     $V' \leftarrow \{u\}$
    for all $v \in (V - V')$ **do**
         $L(v) \leftarrow (v, u)$
    while $V' \neq V$ **do**
        begin
        randomly choose a $u \in (V - V')$ and
        the associated edge e from u to V' is $L(u)$
         $T' \leftarrow T' \cup \{e\}$
         $V' \leftarrow V' \cup \{u\}$
        for all $v \in (V - V')$ **do**
        if $w(L(v)) > w((v, u))$ **then** $L(v) \leftarrow (v, u)$
        end
End.

Fig. 7. Heuristic based on Prim's algorithm to randomly generate a spanning tree

Let T' be a connected subgraph of a spanning tree under construction, and span a subset of vertices $V' \subset V$. Initially V' contains some arbitrary vertex u . At each stage, the label $L(v)$, for each vertex $v \in (V - V')$, records the edge of least weight from v to V' , where $d_{i,j}$ is taken as weight $w((i, j))$. Each $L(v)$ is initialized to edge (u, v). The algorithm updates the $L(v)$ whenever a new vertex u has been added to V'.

At each iterative stage of the algorithm, a new edge e is added to T'. The edge e is selected randomly from L; which is different from Prim's algorithm, where the new edge e is the edge of least weight in L. The algorithm stops when $V = V'$.

We illustrate this algorithm in Figure 8.

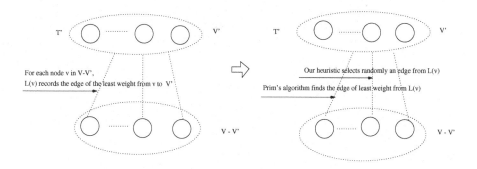

Fig. 8. Illustration of generateSpanningTree()

An initial population is then constructed using this heuristic from randomly generated spanning trees. It keeps random characteristics and the spanning trees in the initial population are of relatively little cost.

5   Algorithm and Results

We have implemented a GA based on the ideas discussed above and tested it on 5 examples we found in the literature.

5.1   Description of the Algorithm

We give an overview of the algorithm in Figure 9.

One execution of the outer "for" loop corresponds to the simulation of one generation. Throught the simulation, the number of individuals $N = |\Phi_c|$ is kept constant (Φ_c is the current population). The algorithm terminates after $maxGen$ generations. The procedures in the algorithm are described as follows:

- *initialPopulation* constructs an initial population Φ_c using the procedure *generateSpanningTree* described in section 4 ;
- *calculFitness* computes the fitness of each individual (spanning tree) as follows: Given a population $\Phi = \{\phi_0, \phi_1, ..., \phi_{(N-1)}\}$, let $C(\phi)$ be the cost of

Procedure geneticAlgorithm()
Begin
```
    intialPopulation(Φc)
    calculFitness(Φc)
    λ = bestOf(Φc)
    for gen = 1 to maxGen do
        begin
            Φn = ∅
            for j = 1 to N/2 do
                begin
                    φ1 = select(Φc),φ2 = select(Φc)
                    crossover(φ1,φ2,ψ1,ψ2) with pc
                    mutation(ψ1), mutation(ψ1) with pm
                    Φn = Φn ∪ {ψ1,ψ2}
                end
            calculFitness(Φn)
            λ = bestOf(Φn)
            Φc = Φn
        end
    output λ
```
End.

Fig. 9. Overview of our GA

individual ϕ, and assume that Φ is sorted so that $C(\phi_0) \geq C(\phi_1) \geq ... \geq C(\phi_{(N-1)})$. The fitness F of ϕ_i is then computed as

$$F(\phi_i) = \frac{2i}{N-1}, \quad i = 0, 1, ..., N-1$$

This fitness computation scheme is called *ranking* (Esbensen 95);
- *bestOf* gives the individual with the highest fitness ;
- *select* selects two parents ϕ_1, ϕ_2 from Φ_c independently of each other, and each parent is selected with a probability proportional to its fitness;
- *crossover* generates two children ψ_1, ψ_2 as described in Section 3;
- *mutation* randomly changes a child as described in Section 3 with a small probability p_m.

5.2   Experimental Results

The algorithm was tested on 5 examples of the OCST problem found in the literature. These examples are 6,12,24 and 35,35 node networks represented as complete graphs, with a distance matrix and a requirement matrix.

The data of 6 and two 35 node networks are given in (Berry *et al.*, 1995, also by http://www.cse.rmit.edu.au/~rdslw/research.html), and the data of 12 and 24 node network is from the thesis of Palmer (1994), but we have not got Palmer's two other examples of 47 and 98 node networks.

Our algorithm is executed 10 times for each example. The parameters for the GA are set as below:

- the crossover probability $p_c = 0.6$;
- the mutation probability $p_m = 0.001$;
- the population size $N = 200$;
- the maximum generation $maxGen = 100$;
- $pathCrossover$, $pathMutation$.

We run our GA on 5 examples and compare to the results with those produced by Berry's GA and Palmer's GA in Table 1.

For the 6 node network, we have the same solution 534 as Berry. For the 35 node network with uniform link distance, the best solution found by Berry is 16915 and the best solution by our GA is 16420; for another 35 node network with non-uniform link distance, the best solution of Berry is 30467 obtained with 250 generations in 50 runs, and by our GA approach the optimal solution 16915 can be reached with 100 generations in 10 runs.

For 12 and 24 node networks, as we could not get the parameters used in Palmer's GA, we compare only on the result quality. We have the same solution 6.857×10^6 as Palmer for the 12 node network, and a better solution 2.173×10^6 for the 24 node network than 2.183×10^6 of Palmer.

Table 1. Comparison of solution quality for GAs

n	GA of Berry GA of Palmer	our GA
6	534	534* (after 1 generation)
35 (non-uniform distance)	30467	16915 * (after 31 generations)
35 (uniform distance)	16915	16420 (after 39 generations)
12	6.857×10^6	6.857×10^6 (after 13 generations)
24	2.183×10^6	2.173×10^6 (after 14 generations)

We give the result of 10 consecutive runs for the instance of 35 node with non-uniform distance in Table 2. From this table, we see that our GA generates initial solutions (about 20000) near the optimal solution (16915) (while a completely random initial solution is about 200000), and then the crossovers and mutations allows to get the optimal solution.

Table 2. Ten consecutive runs for the 35 node network with non-uniform distance

	1	2	3	4	5	6	7	8	9	10
initial solution	22157	25867	25529	24033	24297	23647	24507	21017	24887	26323
final solution	16915*	17029	18771	16915*	17653	17057	17181	17625	17181	17181

An entry marked by a * corresponding an optimal solution.

6 Conclusion and Future Work

In this paper, we present a new GA to solve the OCST problem in which tree structure itself is taken as chromosome, crossovers and mutations directly manipulate on trees, and an initial population is generated randomly by a simple heuristic based on Prim's algorithm. The computational results show that the proposed GA is simple and effective to get the optimal or near-optimal solution to the OCST problem.

Our work is an effort to overcome the limitation of traditional linear chromosome by allowing GAs directly work on the natural structure of problems under consideration. We also try to improve traditional uniform initialization procedure by generating randomly an initial population of "good quality".

In the future, we will test our GA on more examples and improve the implementation of crossovers and mutations. We will also adapt our GA to other problems.

Acknowledgments. We thank Berry, Murtagh, Sugden and McMahon for creating a pageweb on the OCST problem from which we get their examples. We thank Palmer for providing us his work and his some examples in his thesis. We thank also anonymous referees for their comments which helped improve this paper.

References

1995. Abuali F.N., Wainwright R.L., Schoenefeld D.A. 1995: Determinant factorization: A new encoding scheme for spanning trees applied to the probabilistic minimum spanning trees problem. Proc. 6th Int. Conf. on Genetic Algorithms (ICGA'95), University of Pittsburgh, USA, 470-477

1995. Berry L., Mutagh B., Sugden S., McMahon G. 1995: Application of genetic-based algorithm for optimal design of tree-structured communication networks. In Proceedings of the Regional Teletraffic Engineering Conference of the International Teletraffic Congress, South Africa, 361-370.

1998. Crecenzi P., Kann V. 1998: A compendium of NP optimization problems. Available online at http://www.nada.kth.se/theory/compendium/, Aug. 1998

1995. Esbensen H. 1995: Computing near-optimal solutions to the steiner problem in a graph using a genetic algorithm. Networks, vol. 26 (1995) 173-185.

1985. Gibbons A. 1985: Algorithmic graph theory. Cambridge University Press, New York.

1989. Goldberg D. E. 1989: Genetic algorithms in search, optimization and machine learning. Addison-Wesley (Reading, Mass).

1974. Hu T. C. 1974: Optimum communication spanning trees. SIAM J. on Computing, 188-195.

1978. Johnson D.S., Lenstra J.K., Rinnooy Kan A.H.G. 1978: The complexity of the network design problem. Networks, vol. 8(1978) 279-285

1995. Julstrom B.A. 1993: A genetic algorithm for the rectilinear steiner problem. Proc. 15th Int. Conf. on Genetic Algorithms, University of Illinois at Urbana-Champaign (S.Forrest, Ed.). Morgan Kaufmann, San Mateo, CA, (1993) 474-479.

1997. Kallel L., Schoenauer M. 1997: Alternative random initialization in genetic algorithms. in Proceedings of the Seventh International Conference on Genetic Algorithms, Morgan Kaufmann, 268-275.

1994. Palmer C. C. 1994: An approach to a problem in network design using genetic algorithms. PhD Thesis, Polytechnic University, Computer Science Department, Brookly, NewYork.

1995. Palmer C. C., Kershenbaum A. 1995: An approach to a problem in network design using genetic algorithms. Networks, vol. 26 (1995) 151-163.

1998. Peleg D., Reshef E. Deterministic polylog approximation for minimum communication spanning trees. Proc. 25th Int. Colloquium on Automata, Languages and Programming, Lecture Notes in Comput. Sci. 1443, Springer-Verlag, 670-679.

1995. Radcliffe N.J., Surry P. D. 1995: Fundamental limitation on search algorithms: Evolutionary computing in perspective. In "Computing Science Today: Recent Trends and Developments", Lecture Notes in Computer Science, Volume 1000, page 275-291, (Ed: Jan van Leeuwen, Springer-Verlag).

1998. Sinclair M.C. 1998: Minimum cost routing and wave-length allocation using a genetic-algorithm/heuristic hybrid approach. Proc. 6th IEE Conf. on Telecommunications, Edinburgh, UK, 66-71.

1996. Surry P. D., Radcliffe N.J. 1996: Inoculation to initialise evolutionary search. In "Evolutionary Computing: AISB Workshop", (Ed: T. Fogarty, Springer-Verlag).

1995. Wolpert D. H., Macready W. G. 1995: No free lunch theorems for search. Technical Report SFI-TR-95-02-010, Santa Fe Institute.

Studies on Dynamics in the Classical Iterated Prisoner's Dilemma with Few Strategies
Is There Any Chaos in the Pure Dilemma ?

Philippe Mathieu, Bruno Beaufils, and Jean-Paul Delahaye

Laboratoire d'Informatique Fondamentale de Lille (UPRESA 8022 CNRS)
Université des Sciences et Technologies de Lille – UFR d'IEEA – Bât. M3
59655 Villeneuve d'Ascq Cedex – FRANCE
`mathieu@lifl.fr, beaufils@lifl.fr, delahaye@lifl.fr`

Abstract. In this paper we study Classical Iterated Prisoner's Dilemma (CIPD) dynamics of pure strategies in a *discrete* and *determinist* simulation context. We show that, in some very rare cases, they are not quiet and ordered. We propose a classification of ecological evolutions into categories which represent complex dynamics, such as oscillatory movements. We also show that those *simulations* are very sensitive to initial conditions. These experimentations could call into question classical conclusions about interest of cooperation between entities playing CIPD. They may be used to explain why it is not true that cooperation is always the convergent phenomenon observed in life.

1 The Classical Iterated Prisoner's Dilemma

When they were at the RAND Corp. Merill M. FLOOD and Melvin DRESHER tried to introduce some kind of *irrationality* in Game Theory, [13,11]. They introduced a simple two person, non zero-sum, non cooperative and simultaneous game, [6]. This game, very simple to describe, covers a large scale of *real life* situations and seems to catch the definition of conflicts of interests. Thus a lot of different kind of work has been done on it, involving not only mathematicians, but also social, zoological, biological as well as computer scientists. The game becomes the most used theoretical model for studying the cooperation and the evolution of cooperation in population of agents.

The game, called the Prisoner's Dilemma, could be described very simply in the following way: let us meet two *artificial agents* having two choices (two *strategies*):

- COOPERATE, let us write **C**, and say to be *nice*
- DEFECT, let us write **D**, and say to be *naughty*

The payoff of each player depends on the moves played by the two agents. Tab. 1 names the score of each case.

The dilemma comes when exploitation of one by the other (T) is better payed than cooperation between the two (R), which itself pays more than a case where

C. Fonlupt et al. (Eds.): AE'99, LNCS 1829, pp. 177–190, 2000.
© Springer-Verlag Berlin Heidelberg 2000

Table 1. CIPD payoff matrix. Row player score are given first.

	Cooperate	Defect
Cooperate	$R = 3$, $R = 3$ *Reward* for mutual cooperation	$S = 0$, $T = 5$ *Sucker*'s payoff *Temptation* to defect
Defect	$T = 5$, $S = 0$ *Temptation* to defect *Sucker*'s payoff	$P = 1$, $P = 1$ *Punishment* for mutual defection

the two tried two exploite each other (P), which finally is a better choice than to be exploited (S). This can be formalised as:

$$S < P < R < T \tag{1}$$

The dilemma stands on the fact that individual interest (NASH equilibrium) differs from collective one (PARETO issues).

The one shot game, involving rational agents and pure strategies, is solved in Game Theory by the NASH equilibrium, which is to always betray its partner: choosing the **D** strategy. In an iterated version players meet each other more than one time. The payoff of an agent is then simply the sum of each of its meeting's payoff. The game is called the Classical Iterated Prisoner's Dilemma (CIPD).

In order to favour cooperation over defection the following constraint is added:

$$S + T < 2R \tag{2}$$

A classical choice of payoff values, mainly introduced by [1], is given in Tab. 1. With such an iterated game what the opponent did on previous moves clearly influences the way an agent will play on next ones. It is then possible to define more strategies than with the one shot version. Let us define some simple ones, some of which will be used in our experimentations :

all_c corresponds to the **C** strategy of the one shot game: it always plays **C**

ll_d corresponds to the **D** strategy of the one shot game: it always plays **D**

tit_for_tat cooperates on the first move then plays opponent's previous move.

per_cd plays periodically C then D, let us note **(CD)***

per_ddc plays **DDC***

per_ccd plays **CCD***

per_ccccd plays **CCCCD***

soft_majo plays opponent's majority move, cooperating in case of equality

prober plays **(DCC)**, then it defects in all other move if opponent has cooperated in move 2 and 3, and plays as **tit_for_tat** in other cases

The main problem in CIPD study is not only to find *good* strategies, but also to understand the dynamic of populations of agents using fixed strategies.

2   Ecological Tournaments and Other Simulations

Two kinds of experimentation are used in litterature to evaluate strategies for the CIPD:

- The basic one, is to make a two-by-two round robin tournament between strategies. The payoff of each one would be the total sum of each iterated game[1]. A ranking could then be computed according to the score of each strategy.
  The higher a strategy is ranked, the better it is.
  As shown in previous work, [4], some cycles between strategies may be found (**A** better than **B**, which is better than **C** which is better than **A**), the order created by this method cannot be considered as total.
- The second kind of experimentation is a kind of imitation of the natural selection process, and is closely related to population dynamics, but in a completely discrete context. Let us consider a population of N players, each one adopting a particular strategy. At the beginning we consider that each strategy is equally represented in the population. Then a tournament is made, and good strategies are favoured, whereas bad ones are disadvantaged, by a proportional population redistribution. This redistribution process, also called a generation, is repeated until an eventual population stabilisation, i.e. no changes between two generations.
  A good strategy is then a strategy which stays alive in the population for the longest possible time, and in the biggest possible proportion. This kind of evaluation quotes the *robustness* of strategies.
  This looks like prey-predator model, but is not. The number of involved species is not limited to two, interactions between, or into, species are much more complex, and global population is fixed. Once a population has disappeared it has no way to reappear: there is no stochastic perturbations nor in population distribution, nor in strategies description.

Let us detail the computation method for ecological evolution involving 3 strategies. This will be used to compute results detailed later in this paper.
Suppose that, initially, the population is composed of 3 strategies **A**, **B**, **C**.
At generation n each strategy is represented by a certain number of individuals: $W_n(\mathbf{A})$, using **A**, $W_n(\mathbf{B})$ using **B** and $W_n(\mathbf{C})$ using **C**.
The payoff matrix of two-by-two meeting between **A**, **B** and **C** is computed and is thus known. $V(\mathbf{A}|\mathbf{B})$ is the score of **A** when it meets **B**, etc.
Let us suppose that the total size of the population is fixed and constant. Let us note it Π.

$$\forall i \in [1, \infty[, \Pi = W_i(\mathbf{A}) + W_i(\mathbf{B}) + W_i(\mathbf{C})$$

The computation of the score (distributed points) of a player using a fixed strategy, at generation n is then :

---

[1] In our experiments every meeting has the same length (1000 moves), but strategies can't guess it.

$$g_n(A) = W_n(\mathbf{A})V(\mathbf{A}|\mathbf{A}) + W_n(\mathbf{B})V(\mathbf{A}|\mathbf{B}) + W_n(\mathbf{C})V(\mathbf{A}|\mathbf{C}) - V(\mathbf{A}|\mathbf{A})$$
$$g_n(B) = W_n(\mathbf{A})V(\mathbf{B}|\mathbf{A}) + W_n(\mathbf{B})V(\mathbf{B}|\mathbf{B}) + W_n(\mathbf{C})V(\mathbf{B}|\mathbf{C}) - V(\mathbf{B}|\mathbf{B})$$
$$g_n(C) = W_n(\mathbf{A})V(\mathbf{C}|\mathbf{A}) + W_n(\mathbf{B})V(\mathbf{C}|\mathbf{B}) + W_n(\mathbf{C})V(\mathbf{C}|\mathbf{C}) - V(\mathbf{C}|\mathbf{C})$$

Let us quote that because of the substractions, computation of g cannot be simplified. The total points distributed to all involved strategies are :

$$t_n = W_n(\mathbf{A})g_n(\mathbf{A}) + W_n(\mathbf{B})g_n(\mathbf{B}) + W_n(\mathbf{C})g_n(\mathbf{C})$$

The size of each sub-population at generation $n + 1$ is finally:

$$W_{n+1}(\mathbf{A}) = \frac{\Pi W_n(\mathbf{A})g_n(A)}{t(n)} \tag{3}$$

$$W_{n+1}(\mathbf{B}) = \frac{\Pi W_n(\mathbf{B})g_n(B)}{t(n)} \tag{4}$$

$$W_{n+1}(\mathbf{C}) = \frac{\Pi W_n(\mathbf{C})g_n(C)}{t(n)} \tag{5}$$

All divisions being rounded to the nearest lower integer.
Classical results on the problem, which have been emphasized by AXELROD in [1], show that to be good a strategy has to:

- be nice, i.e. not be the first to defect[2]
- be reactive
- forgive
- not be too clever, i.e. to be simple in order to be understood by its opponent

The well-known **tit_for_tat** strategy which satisfies all those criteria, has, since AXELROD's book, been considered to be one of the *best* strategies not only for cooperation but also for evolution of cooperation.
It is widely accepted that cooperation seems to be the more general adopted behavior with this model. However it is also clear that there is a gap between those classical results and what appears in the life-as-it-is. There are not only *nice* people. Cooperation, as choice of collective fitness against individual one, is not the emergent stable behavior in all ecosystems.
The model carries those contradictions in some way. For instance, it is possible for a *naughty* strategy population to evoluate better than a nice one, and thus to win an ecological simulation, as seen in the example set up in Fig. 1.
Cooperation has been thought as a *global convergency point* of (artificial) living systems. Definition of ecological evolution, i.e. without any mutation of individuals ; simplicity, and small size, of studied population set ; limitation of computing power, may be explanations of the differences found between formal results and practical constatations.

---

[2] whereas *naughty* strategies defects spontaneously at least one time

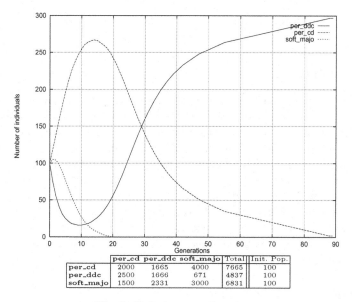

	per_cd	per_ddc	soft_majo	Total	Init. Pop.
per_cd	2000	1665	4000	7665	100
per_ddc	2500	1666	671	4837	100
soft_majo	1500	2331	3000	6831	100

Fig. 1. Defectors may be strong

We think that the *simplicity* criteria in the strategy definition is not good, [5], and we have thus introduced a strategy called **gradual**[3], which illustrates our point of view, [2]. We have evaluated it in large environments, [3].

The strategy's point of view is however not the only one to be taken into account when trying to understand evolution of cooperation. Population dynamics play a major role too. So we think it is important to understand it well. In particular it is important to know if *chaotic* dynamics are possible.

It may seem easy to find a particular round robin payoff matrix which could imply complex dynamics in ecological evolutions. But verifying such a matrix corresponds to some strategy sets, and defining those strategies, is harder. The only useful way to study dynamics is then to make systematic exploration.

Previous works, [10,9,8], have shown that with stochastic strategies, or under evolutionary conditions, oscillations in evolution of population could often be found. In those particular cases, the stochastic, i.e. non deterministic, element could be one of the main explanations to the oscillatory dynamics observed. Other analytic results are widely spread in the case of population dynamics computed in a not discrete way, see for instance [7].

As we think that to understand complex cases one has to first understand well the behavior of simple, we studied the simplest ones. Thus all strategies we will

---

[3] **gradual** cooperates on the first move, then after the first opponent's defection defects one time, and cooperates two times, after the second opponent's defection defects two times and cooperates two time, ..., after the nth opponent's defection defects n times and cooperates two times

use in this paper, unlike in [10], are *pure* in the Game Theory meaning. That is they are deterministic.

3   Unexpected Dynamics

In most cases, ecological evolutions look like monotonous convergence, which means that population's evolution curves are always increasing or decreasing. The ranking seems to be determined after few steps. In some cases, however, complex oscillations can be observed. For instance, one can obtain oscillatory movements, which could be attenuated, increasing or periodic. We have undertaken to systematically seek such cases, and we have found some which seems to be at some kind of *"edge of chaos"*. It is easier to find oscillations with many strategies but we will see that such oscillations can be found with few ones. In this paper we have chosen to present only situations with three pure strategies involved. After having analysed thousands of evolutions, we propose to classify the phenomena observed in five groups.

3.1   Monotonous Convergence

The first of the five groups corresponding to a great majority of cases (99% in our experiments) and which is often thought to be the only one, is that of monotonous convergence. Population's size after such evolution (no or little change in the movement) stabilizes itself completely (see Fig. 2).

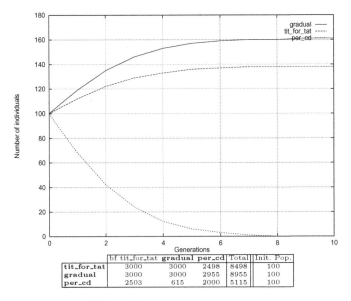

	bf tit_for_tat	gradual	per_cd	Total	Init. Pop.
tit_for_tat	3000	3000	2498	8498	100
gradual	3000	3000	2955	8955	100
per_cd	2503	615	2000	5115	100

Fig. 2. Monotonous convergence

3.2   Attenuated Oscillatory Movements

The second case is the attenuated oscillatory movement one. The size of the population oscillates with a decreasing amplitude, which leads at the end of the evolution, as in the first case, to a population stabilization, but this time after many reversals.

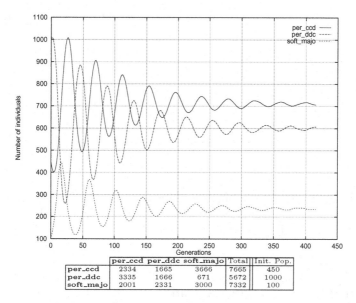

	per_ccd	per_ddc	soft_majo	Total	Init. Pop.
per_ccd	2334	1665	3666	7665	450
per_ddc	3335	1666	671	5672	1000
soft_majo	2001	2331	3000	7332	100

Fig. 3. Attenuated oscillatory movements

Fig. 3 illustrates this case. Three populations of strategies **per_ccd**, **per_ddc** and **soft_majo** are conflicting themselves with many oscillations during the first 100 generations and then, gradually, find an equilibrium which is reached with generation 420 from which no more modification occurs.

3.3   Periodic Movements

The third case is the periodic movements one. Population size of the strategies after a potential phase of hesitation recurringly evolves reproducing after several generations the same combination, without stabilizing (see Fig. 4). On this example, population size comes at the same point every 37 generation. The oscillation is never stabilized.

It seems that such phenomena involves strategies which made a cycle in tournament : **A** is better than **B**, which is better than **C** which is better than **A**. The nature of those relation may be an explanation of those periodic movements.

Such a phenomenon was recently quoted in [12] in the living world and relates to populations of lizards. Even if it is not sure that those results could be applied

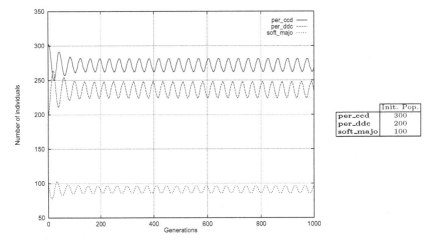

Fig. 4. Periodic movements

in the CIPD model, it is useful to notice the coincidence between our three strategies population periodic movements on the one hand, and the real world of terrestrial life on the other hand.

3.4 Increasing Oscillations

The fourth case, is the increasing oscillations with rupture. The case represented by Fig. 5 is similar to the precedent one, except that now oscillations are growing. It leads at the end to a *break*.
The break is done in profit of **per_ddc** which remains alone, after many oscillations. This kind of dynamics shows that violent oscillations can allow the survival of non-cooperative strategies which benefit of the general disorder.

3.5 Disordered Oscillations

The fifth kind of dynamics gathers the cases which don't get into the fourth previous ones. Movements seem disordered. In our experiments these disordered movements do not last long, therefore we hesitate to use the *chaos* term. On Fig. 6, after a strong instability during 250 generations where each of the three strategies comes very close to death, an equilibrium point is reached.

4 Sensitivity to Initial Conditions

In order to try to complete our opinion on the *chaotic aspect* of those dynamics we studied its sensitivity to initial conditions. We found that very small variations of initial parameters could imply important changes in the phenomenom observed.

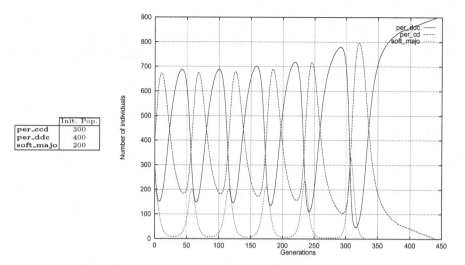

Fig. 5. Increasing oscillations

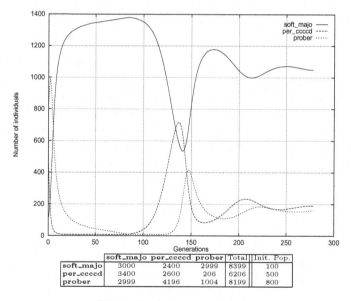

Fig. 6. Disordered oscillations

4.1   Sensitivity to Population's Size

The transition from a periodic movement to a monotonous one can be made when the initial size of the population varies from one unit.

In the first experiment of Fig. 7, the CIPD parameters are the classical ones (the one represented on Tab. 1, T=5, R=3, P=1 and S=0), each match is 1000 meeting long, there are 300 agents using **per_ccd**, 100 using **soft_majo** and 244 using **per_ddc**. Populations evoluate in a periodic movement.

If only one **per_ddc** agent, which is a *naughty* one, is added then the evolution is a monotonous convergence.

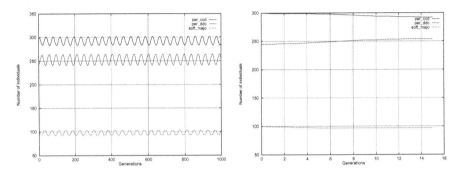

Fig. 7. Sensitivity of dynamics to population's size. All parameters are identical except for the initial size of **per_ddc** which is 244 on the left and 245 on the right

Variation of one unit in the initial population of a strategy can also change the winner of the ecological evolution.

In the experiment of the Fig. 8, the conditions are the same as for the previous ones, except for the size of population. There are 100 **per_ddc**, 159 **soft_majo**, and 100 **per_cd**. The winner is **per_ddc**. If only one **soft_majo** is added, then the winner is **per_cd**. It could be noticed that the modified strategy **never** wins in any cases.

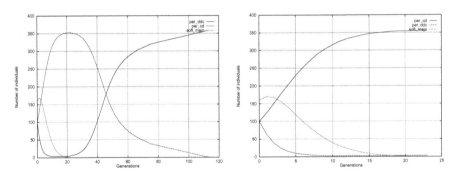

Fig. 8. Sensitivity of winner to population's size. All parameters are identical except for the initial size of **soft_majo** which is 159 on the left and 160 on the right.

4.2   Sensitivity to Game Length

A change of the kind of dynamic can be created by the variation of the length of game (number of iteration of the Prisoner's Dilemma), which is fixed but unknown by strategies.

In the experiment of Fig. 9 the CIPD is used (5,3,1,0). There are 300 **per_ccd**, 100 **soft_majo**, and 244 **per_ddc**. When the game lasts 7 moves the dynamic is a periodic movement and becomes an attenuated oscillary one when the game is 6 moves long.

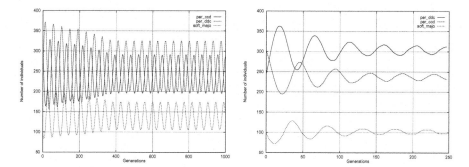

Fig. 9. Sensitivity to game length. All parameters are identical except for the game length which is 7 moves on the left and 6 on the right.

4.3   Sensitivity to CIPD Payoff

A change in the Prisoner's Dilemma payoff matrix, with respect to inequations 1 and 2, may change the kind of dynamic.

In the experiment of Fig. 10 there are 300 **per_ccd**, 100 **soft_majo** and 244 **per_ddc**. Games last 1000 moves. R, P, and S are the same as in the classical choice of the Tab. 1, but T=4.6. The dynamic is an increasing oscillation movement. When T=4.7 then it comes to a periodic one.

4.4   Sensitivity to Repartition Computation Method

Dynamics can change with the individual repartition method used in ecological evolution between two generations (rounding computation method).

In the experiment of Fig. 11 the CIPD parameters are the classical one (5,3,1,0), games are 1000 moves long, there are 300 **per_ccd**, 100 **soft_majo** and 200 **per_ddc**. If we round the number of individuals as in equations 3, 4 and 5 the dynamic is a periodic movement, whereas if we use real values (populations are no more discrete), it comes to an attenuated oscillation one.

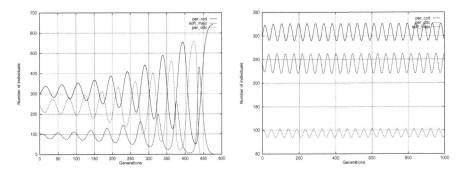

Fig. 10. Sensitivity to CIPD payoff. All parameters are identical except that T=4.6 on the left and T=4.7 on the right.

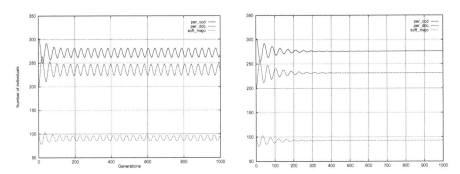

Fig. 11. Sensitivity to repartition computation method. All parameters are identical except that repartition on the left is done by rounding and uses real value on the right.

Another proof of this sensitivity to rounding computation method is that when multiplying all proportions by a constant factor, dynamic changes.

In the experiment of Fig. 12, the CIPD parameters are the classical ones (5,3,1,0), games are 1000 moves long, there are 450 **per_ccd**, 100 **soft_majo** and 1000 **per_ddc**. Dynamic is an attenuated oscillating movement. If all populations are divided by 10, it becomes an increasing one.

We also note that small modifications in the composition of strategies sometimes involve the disappearance of the oscillation, or change the winner of an increasing oscillation. In most of the cases the shape of the curves is different.

5   Conclusion

In some very rare and particular cases, ecological evolution comes to disordered population dynamics, which we classify into 5 categories. In the periodic cases, with attenuated or increasing oscillations, there are always some *naughty* strategies involved, as expected. Sometimes, during the break of an attenuated

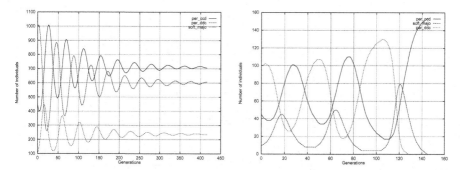

Fig. 12. Sensitivity to repartition computation method. All parameters are identical except that populations on the right are divided by 10.

oscillation the winner is a *naughty* one. It seems that disorder gives more chance to *not nice* strategies and is unfavourable to cooperative ones.

Instability to initial conditions, and the fact that the winner is not always a *nice* one, makes the way the population evolves almost unpredictable if using only round robin tournament results and ecological evolution equations.

It is possible that with highly complex strategies, what seems exceptionnal here becomes more frequent. If it is the case, this would mean that, unlike the most accepted interpretation of CIPD, cooperation is not the most frequent attractor state, when agents with complex behavior are involved.

Social relations would then be unstable by nature because of oscillatory dynamics leading to break, which benefits to aggressive strategies.

Simulation software with many strategies is already available for all plateforms, through our web site at **http://www.lifl.fr/IPD** or by anonymous ftp at **ftp.lifl.fr** in **pub/projects/IPD**.

References

[1] R. Axelrod. *The Evolution of Cooperation*. Basic Books, New York, USA, 1984.

[2] B. Beaufils, J.P. Delahaye, and P. Mathieu. Our Meeting with Gradual, A Good Strategy for the Iterated Prisoner's Dilemma. In Christopher G. Langton and Katsunori Shimohara, editors, *Proceedings of the Fifth International Workshop on the Synthesis and Simulation of Living Systems*, pages 202–209, Cambridge, MA, USA, 1996. The MIT Press/Bradford Books. Artificial Life 5, Nara, Japan, May 16-18 1996.

[3] Bruno Beaufils, Jean-Paul Delahaye, and Philippe Mathieu. Complete classes of strategies for the classical iterated prisoner's dilemma. In V. W. Porto, N. Saravanan, D. Waagen, and A. E. Eiben, editors, *Evolutionnary Programming VII*, volume 1447 of *Lecture Notes in Computer Science*, pages 33–41, Berlin, 1998. Springer-Verlag. Evolutionnary Programing VII, San Diego, CA, USA, March 25-27, 1998.

[4] J.P. Delahaye and P. Mathieu. Expriences sur le dilemme itr des prisonniers. Publication Interne IT-233, Laboratoire d'Informatique Fondamentale de Lille, Lille, France, 1992.

[5] J.P. Delahaye and P. Mathieu. Complex Strategies in the Iterated Prisoner's Dilemma. In A. Albert, editor, *Chaos and Society*, volume 29 of *Frontiers in Artificial Intelligence and Applications*, pages 283–292, Amsterdam, Netherlands, 1995. Universit du Qubec Hull, Canada, IOS Press/Presses de l'Universit du Qubec. Chaos & Society 1994, Trois Rivires, Canada, June 1-2 1994.

[6] Merill M. Flood. Some experimental games. Research memorandum RM-789-1-PR, RAND Corporation, Santa-Monica, CA, USA, June 1952.

[7] Josef Hofbauer and Karl Sigmund. *Evolutionary Games and Population Dynamics*. Cambridge University Press, Cambridge, UK, 1998.

[8] K. Lindgren. Evolutionary Phenomena in Simple Dynamics. In Christopher G. Langton, Charles Taylor, J. Doyne Farmer, and Steen Rasmussen, editors, *Artificial Life II: Proceedings of the Second Interdisciplinary Workshop on the Synthesis and Simulation of Living Systems*, volume 10 of *Santa Fe Institute Studies in the Sciences of Complexity*, pages 295–312, Reading, MA, USA, 1992. Addisson-Wesley Publishing Company. Artificial Life 2, Santa Fe, USA, February 1990.

[9] John H. Nachbar. Evolution in the finitely repeated prisoner's dilemma. *Journal of Economic Behavior and Organization*, 19:307–326, 1992.

[10] M. Nowak and K. Sigmund. Oscillations in the Evolution of Reciprocity. *Journal of Theoretical Biology*, 137:21–26, 1989.

[11] W. Poundstone. *Prisoner's Dilemma : John von Neumann, Game Theory, and the Puzzle of the Bomb*. Oxford University Press, Oxford, UK, 1993.

[12] J. Maynard Smith. The Games the Lizards Play. *Nature*, 380:198–199, March 1996.

[13] John von Neumann and Oskar Morgenstern. *Theory of Games and Economics Behavior*. Princeton University Press, Princeton, NJ, USA, 1944.

An Adaptive Agent Model for Generator Company Bidding in the UK Power Pool

A. J. Bagnall and G. D. Smith

School of Information Systems, University of East Anglia, Norwich, England

Abstract. This paper describes an autonomous adaptive agent model of the UK market in electricity, where the agents represent electricity generating companies. We briefly describe the UK market in electricity generation, then detail the simplifications we have made. Our current model consists of a single adaptive agent bidding against several non-adaptive agents. The adaptive agent uses a hierarchical agent structure with two Learning Classifier Systems to evolve market bidding rules to meet two objectives. We detail how the agent interacts with its environment, the particular problems this environment presents to the agent and the agent and classifier architectures we used in our experiments. We present the results and conclude that using our structure can improve performance.

1 Introduction

This paper describes work on a project supported by the National Grid Company, UK (NGC), examining the use of Artificial Adaptive Agents that use Learning Classifier Systems (LCS) to model electricity generator strategies within the UK Power Pool. The generation and supply of electricity for the UK was privatised on March 31st 1990. Electricity is produced by generating companies and sold via the Electricity Pool, administered by the NGC, to electricity suppliers. As part of this process, each day generators make a complex set of bids to provide power for a fixed period the following day. We describe in Section 2 the salient features of the bidding process which are relevant to our simplified model. A fuller description of the market can be found in [1].

The simplified model consists of a single adaptive agent attempting to develop bidding strategies to meet the objectives of not making a loss and maximizing profit while competing against non-adaptive agents (i.e. agents with static bidding rules designed to model real world strategies). This problem can be summarised as a one step stimulus/response situation with the agent having to choose one of 32 possible actions for any one of 1024 environments in order to meet one of the two objectives. We have attempted to model the market in a way that allows analysis of performance of the adaptive agent while maintaining some important features of the actual market. The profit function is real valued and this, coupled with the comparatively large action space and the unequal probabilities of the environments, make this a problem hard for a LCS to solve.

C. Fonlupt et al. (Eds.): AE'99, LNCS 1829, pp. 191–203, 2000.

In Section 3 we attempt to justify our reasons for designing the problem in such a way by referring to the real world problem. In order to gain an adequate level of performance in this environment we have developed a complex agent structure which is described in Section 4. The agent utilizes two LCSs to meet its objectives and the specifics of these are given in Section 5. Some results and rules used to obtain these results are given in Section 6.

2   The Simplified Market Model

A generating company has one or more units that can generate electricity on to the National Grid. There are approximately 200 generating units regularly competing to generate power. Each day a generating company must produce an *offer bid*, or just bid, for every unit it owns. A bid gives details of the conditions under which the generating company is willing to have a unit generate. A bid consists of up to 8 price parameters and a host of technical parameters. The NGC collects these bids and uses them to form a schedule of generation for each unit. The simplified model consists of 21 units, each controlled by an agent, i.e. each generating company has a single unit. Of these 21 agents, one agent is an autonomous adaptive agent, the other 20 agents are non-adaptive agents whose bidding is controlled by deterministic rules. A bid for a unit is a single quantity representing the Table A bid price (a summary measure of the bid parameters, see [1]).

An iteration of our simplified model, representing a single day, consists of the following steps:

• **Generate Environment**. Generate the relevant environmental information for the following day and pass this information to the agents. Details of the environment are given in Section 3.

• **Bidding**. Every unit produces a bid for the next day's generation. These bids are collected and passed to the scheduling unit.

• **Unconstrained Scheduling**. The scheduling unit decides, with the objective of minimizing cost, which units should be allowed to generate in order to meet forecast demand for the following day. The forecast demand curve consists of 48 half hourly estimates of the amount of electricity that will be required. This type of scheduling problem is commonly called the *unit commitment problem* [2]. The unconstrained schedule consists of a generation profile for every unit submitting a bid. We use a simple *merit order loading* method to form the unconstrained schedule. The merit order loading method involves simply ranking the bids by price then loading each half hour time slot in ascending order of price until demand is met.

• **Constrained Scheduling**. The constrained schedule is formed to satisfy constraints on the actual network of wires. If a unit is constrained it may be constrained on (required to generate whether it was scheduled to or not) or constrained off (forced to not generate). For the constraining process in our simplified scenario the units are separated into three fixed groups of seven units. There can be a maximum of one group constrained on and one group constrained off at any one time. If a group is constrained on then there is a minimum level of power which must be generated by those units in the group for each half hour time slot. The scheduler increases the generation of the units in the group up to each unit's maximum, in order of price (cheapest first), until the constraint is met. Conversely, if a group is constrained off there is a maximum half hourly level of generation for those in the group, and units are scheduled off in order of price.

• **Settlement.** Payment for power generated is based on two things: the unconstrained schedule and the *capacity premium*. The unconstrained schedule is used to form the *System Marginal Price* (SMP). The SMP consists of 48 prices in pounds per MWh, and is set as the bid of the marginal generator for each period (i.e. the most expensive unit scheduled to generate). The capacity premium is a function of what is called the *loss of load probability* and is added to the SMP to form the *Pool Purchase Price* (PPP), the amount that unconstrained units are paid for the power they generate. So, if a unit is not the marginal generator but is scheduled to run it still receives the same payment as the marginal unit for power it generates. The payment rules for units constrained on or off are different. If a unit is constrained off it is paid at *(SMP- bid price)* for the power it was not allowed to produce despite being scheduled to do so. A constrained on unit is paid at bid price for the power it was constrained to produce.

3   The Problem Facing The Adaptive Agent

For our model we simulate three of the key factors in the real world that influence the way generators bid: Constraint information, demand level, and capacity premiums. The most obvious observed real world strategies depend on these factors, and this allows us to construct our profit function to reflect this. For example, an obvious strategy when constrained on is to bid as high as possible, since the constrained unit is paid at bid price. There are many examples in the electricity market of generating companies exploiting constraints. The agent detector maps the constraint information onto the first six bits of the environment message. Bits one and two represent the constrained on group (00 if no group is constrained on) bits three and four represent the constrained off group and bits five and six represent the 4 possible levels of constraint allowed in this model. The Forecast Demand data is highly seasonal and the profile of the demand curve and the level of demand differs considerably between weekday and weekend. We therefore categorize the demand curve as either typical to weekend/weekday winter level (bit 7 of the environmental message), or weekend/weekday summer level (bit 8). We assume that the generators know, or at least are able to estimate, the capacity premiums prior to making a bid. As with demand we need to characterize the graphs to limit the quantity of information passed to the agent, and through observation of the real world data we identified four distinct types of capacity premium curves determined by bits 9-10 of the message. The non-adaptive agents are loosely modeled as one of five types of units that bid : Nuclear, Gas, Coal, Oil or Gas Turbine (GT). There are five Nuclear, five Gas, five Coal units and three Oil units. Each of these has a capacity of 3000MW.

There are two GT units with capacity 500 MW. The setting of the strategy of the non-adaptive units has an obvious effect on the profit matrix of the adaptive unit, and so may be adapted to make the task for the LCS easier or harder. The bidding strategies of the units have been designed to reflect their station type, but also to present the adaptive agent with certain problems in the profit function.

The model presents the agent with an environment that consists of 1024 possible states and 32 different possible actions, each with an associated profit. An action is an integer between 0 and 31. The effector copies the agent action to the agent bid, a Table A bid price restricted to the values £0 per MWh to £31 per MWh. The profit is calculated from a payment received from the settlement unit for power generated that day. The agent uses this data and the constrained schedule to calculate its profit. We use a linear cost function, and assume generation is at the schedule level throughout the time slot, giving a stepped generation curve. The daily generation cost is simply the sum of the amount generated times the unit generation cost, and the profit is the payment from the settler less the generating costs and the daily fixed cost. The most important features of this profit function are:

• Some of the most profitable environments (such as maximum capacity premium) do not occur very often. This means we will have to be careful to maintain niching within our rule set.

• There are large differences in the profit obtainable and the agent has no explicit way of knowing how well it could have done in any particular environment. This means good rules may still have a high standard deviation and conversely poor rules may have a high average profit.

• The profit function has areas of irregularity between close environments, meaning more specific rules will be needed to do well in some environments.

These issues are some of the factors that have influenced the agent structure we have adopted, which is described in the next section.

4   Agent Structure

The objective of the agent is to develop a strategy that optimizes its profit within the specified environment. Maximizing the profit has proved to be a non-trivial problem for an agent using a LCS (see Section 7). We have, however, continued to favour an LCS based architecture over the possible alternative methods.

The task of meeting each objective is essentially a deterministic Markov decision problem, and as such a temporal difference algorithm such as Q-learning would probably provide a more efficient method of finding an optimum strategy over the environment/action space. However, the real difficulty in this problem lies in finding good generalizations over the environment space with limited memory capacity. A temporal difference approach requires the explicit storage of a complete model of the environment/action space and makes no generalizations over the environments. We feel that an architecture with efficient generalization capabilities for reinforcement learning problems would provide a more flexible and comprehensible framework for investigating agent interaction in more complex nondeterministic environments.

We also felt that since we have gone to some length to develop a model based on a real world problem, we should be able to interpret the agent's environment generalizations in terms of the market model. This discouraged us from using a neural net approach. A LCS provides rule based system where it is reasonably simple to interpret generalizations made over the environment in terms of broad subsets of market states. For example, we can interpret the rule (01********/25) as meaning "if unit is constrained on bid 25". This kind of interpretation is difficult with neural nets. The attraction of a LCS based system is in the generality of the rule representation and the parallel nature of the action selection and reinforcement mechanisms. However, the cost of these advantages is in a hugely increased space of possible rules and possible overall strategies. This is one of the reasons we have adopted a complex agent structure similar to that used by Dorigo in [3], in that the agent has a high level action-decision controller and learning coordination mechanism. This structure sends signals to and takes input from two classifier systems and, depending on the current agent objective and utilizing some top level long term memory structures which we call *case lists*, uses this information to decide on a final action. The two classifier systems, referred to as CS_A and CS_B, are used to help the agent meet the two objectives we have set it, firstly to develop a rule set that will insure the agent does not make a loss and secondly to discover rules to maximize profit. CS_A starts with a set of randomly generated rules while CS_B starts with no rules at all. The LCS structure is based on Wilson's XCS [4] and is discussed in Section 5.

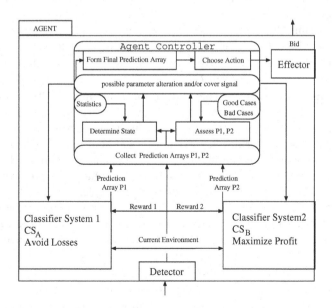

Fig. 1. Top Level Agent Structure

The daily cycle of the agent controller shown in Figure 1 can be summarised as follows:

Collect and Distribute Environment Information

The agent controller receives the environmental information for the current day and the profit data for the previous day. It then calculates its profit and generates a reward for each LCS using two reward functions designed to reflect the two objectives: The first reward function, R_1, maps to zero if the agent made a loss and 100 otherwise. The second reward function is a linear mapping of the profit function to a value between 0 and 1000. The two rewards are passed to the respective LCSs which use them in their reinforcement mechanisms (see Section 5).

Receive Prediction Arrays

CS_A and CS_B then execute their performance systems, both of which result in a *prediction array* [4], called P_1 and P_2 respectively, which are passed back to the controller.

Determine Agent State and Set Agent Parameters

To reflect the priority of the differing agent objectives we say our agent is in either a *defensive* state, when it is primarily concerned with not making a loss, or an *aggressive* state, when it is more interested in maximizing profit. Furthermore, the agent can either be in *explore* mode, where the agent is interested in trying new environment/action pairings, or *exploit* mode, where the agent wants to meet its current objective, be it defensive or aggressive. Thus the agent has four possible states, and these influence how the agent uses the information from the classifier systems to choose an action and how the controller sets the LCS parameters of CS_A and CS_B. We assume that initially the agent's highest priority is learning how to bid so as not to lose money, so the agent starts off in a defensive explore state. The agent decides on which defensive state to adopt and whether to become aggressive using the proportion of profitable bids over the previous year, which we call *prop*.

It remains in a defensive explore state for a maximum of two years or until *prop* rises above 80%. It then switches to defensive exploit state. The agent will switch to aggressive explore when it is happy it has a good fall back position, i.e. when it has a set of rules it is confident will at least make a profit in most of the environments. We characterize this situation by defining the switching point as the agent achieving a *prop* figure of 98%. Once in aggressive mode, the agent uses a different statistic to decide between explore and exploit. It keeps a record of its best annual profit, and if profits drop noticeably below this figure the agent switches to exploit, whereas if this figure has not been improved upon for some time the agent state is set to explore. The effect of the agent state on the CS parameters is described in Section 5.

Determine Whether Cover Signal Needed

A signal may be sent to CS_A invoking the covering mechanism described below. Triggering is invoked when the maximum value of P_1 is below the CoverTrigger threshold (either a constant value 85 if in explore mode or a value equal to *prop* if in exploit). The cover signal is passed to CS_A which then may alter its rule set and pass a new P_1 array back to the controller for reconsideration. The signal consists of the CoverTrigger value and possibly any previously known bad actions from the *bad cases list* (BCL). The agent utilizes two long term memory structures external to the classifiers: the BCL and the *good case list* (GCL). For this problem they have enabled us to both speed up the convergence and improve the quality of results. The purpose of the BCL is to help the agent meet its first objective by recording certain rare environments with few profitable actions, then use this information to alter CS_A and influence the action selection. A case consists of an environmental message and a record of known good (profitable) and bad actions. Both BCL and GCL start empty and have a maximum size of 20 cases. Cases are added to BCL when the agent is in an exploit mode and the agent makes a loss if the decision was based on P_1.

Once the controller has received P_1 and P_2 and determined its current state, it examines the current environmental message and finds if there is a match on the BCL. If there is a match the controller sets any elements of P_1 for which a loss is known to occur to zero before checking for a covering trigger.

The GCL records cases with a rule condition rather than an environmental message (i.e. it can contain wild-cards) and an estimate of the associated R_2 values. The purpose of the GCL is to keep a record of environments observed while in a defensive state that may be of particular interest in increasing profit and helping meet the second objective. The agent classes an environment a candidate good case if the R_2 value received is at least twice the average R_2 value. A candidate is either added to the GCL or an existing case is generalized if it is within one bit of the candidate case. When the agent first makes the change from defensive to aggressive CS_B still contains no rules. The CS_B rule set is generated from the GCL by randomly generalizing the case conditions and selecting a random action.

Form Final Prediction Array

If the agent is in a defensive state it uses P_1 only whereas if it is in an aggressive mood it uses P_2, unless the maximum prediction value of P_2 is less than the maximum value of P_1, in which case it reverts to using P_1. In this way once the agent is in an aggressive state the predictions of CS_A are used as a default position, used only if CS_B has no matches or very weak rules matching the current environment.

Choose Final Action

The action selection method is also dependent on the agent state. If the agent is in an explore mode it chooses the winning action probabilistically. If it is in an exploit mode it chooses the action with the highest value in the final prediction array.

5   Details of CS_A and CS_B

The two LCSs the agent employs have the same performance system and reinforcement mechanisms, but differ in rule discovery methods. CS_A uses reward function R_1 and is required to produce a concise rule set that is good description of $E \times A \rightarrow R_1$, where E is the space of possible environments and A is the action space. CS_A is very similar to Wilson's XCS [4], but has some minor modifications described below. It was found to be difficult to design a LCS to produce a comprehensive description of $E \times A \rightarrow R_2$ when using a relatively small rule set. Instead, we designed the second classifier system to concentrate on the peak areas of R_2, i.e. on areas of the environment space where large profits can be made. These areas are mostly made up of areas where the agent is constrained on or off (messages matching 01******** and **01****** respectively) and where the capacity premium is high (********11) and these environments are generally much rarer than other environments. CS_A starts with 200 randomly generated rules and CS_B can have a maximum of 400 rules, which are initially generated from the GCL when the agent switches to an aggressive state.

Performance Component
A value in the prediction arrays P_1 and P_2 in position a_i is an error weighted average of the prediction values of classifiers in the match set with action a_i. We used error weighting because we found that due to the generally small size of the action sets (due to the relatively small number of rules we use and the large number of possible actions) the fitness parameter could be misleading in certain ways. When a rule is the sole member of the action set its relative accuracy will be 1 whatever the error of the rule, and so the fitnesses of young rules can be somewhat misleading. While this effect is counteracted over time by the GA and further evaluations, it was seen to have a disruptive effect on the speed of convergence to a good strategy for meeting objective 1. The action set is formed at the beginning of the next time step, before the reinforcement stage.

Reinforcement Component
Each rule has a prediction, error and fitness. The prediction and fitness are adjusted using the MAM procedure and the Widrow-Hoff delta rules as described in [4]. For the error parameter we use the standard deviation of the rewards a particular rule has received divided by an estimate of the maximum reward deviation. We see the error parameter as a more long term gauge of performance, so that whereas previous poor rewards will in time have very little impact on current prediction values (depending on the β value used) some record of past mistakes will remain in the error parameter.

Discovery Component of CS_A
The GA used by CS_A differs from that used by XCS only in the fact that it produces a single offspring rather than two, and in the manner of crossing the parents to produce a new action. Since the actions are integers using a bitwise cross can often be very disruptive. Instead an offspring either inherits one of its parents' actions or a new action is chosen by sampling a probability distribution centred around the average of the parents' actions. Mutation on the child's action can also occur and consists of either increasing or decreasing the action by one.

If CS_A receives a cover signal from the controller it forms a new rule with an action not currently present and not on the list of banned actions which may form part of the cover signal (if the current environment is on the BCL). To do this we first examine the rules set to see if there is a very good rule (i.e. a rule with a prediction parameter above the CoverTrigger and small error) that is very close to matching (in that it has only one non-wildcard bit different to the environmental message). If such rules exist one is chosen probabilistically, with weighting in favour of more specific rules, and a new rule is created with the non-matching bit set either to match the environment or set to a wildcard. If no rules fit the matching criteria a matching condition is generated randomly and a random candidate action is chosen. The rule thus created is then merged into the rule set using the normal merge mechanism.

Discovery Component of CS_B

CS_B is not meant to maintain a full covering of possible inputs. Instead it is supposed to maintain a good coverage of the action space for certain areas of the environment space. To achieve this we used a GA that acts on the whole rule set. The GA is triggered periodically and attempts to create one rule for each of the 32 possible actions. The GA only considers rules that have passed the MAM threshold (i.e. have been active at least $1/\beta$ times). For each action two parents are then selected via roulette, with fitness set to the prediction values. If different rules are selected then the condition of the child is create normally via single point crossover. If the same rule is selected twice a new action is chosen for the offspring (either by adding one to the action, subtracting one or randomly choosing) and the child's condition is copied from the parent. Mutation proceeds as before. This GA method is designed to find rules for each action with conditions that offer maximum predicted reward while allowing some copying of conditions from one action to another. It also helps to maintain some diversity by making the creation of duplicates less likely. The new rule replaces a rule chosen panmictically by probabilistically sampling a distribution derived relative to a value that is a function of a rule's prediction, error and age (to slightly favour younger rules in the deletion process). It handles oscillating rules (rules with above average error) by using an operator similar to Dorigo's *mutespec* [5].

Parameter Settings and Changes by the Controller

The agent sets the parameters of the LCS to reflect its current state. Rule discovery is set to be more frequent in explore mode than in exploit mode. While in defensive explore state θ, used in triggering the GA in CS_A, is set to 50. When the agent switches to defensive exploit, θ is increased to 200. When the change to aggressive explore is made, the GA for CS_A is turned off to avoid disrupting our default position. CS_A may still be altered by covering. In aggressive explore the CS_B periodic triggering value is set to 100, whilst in aggressive exploit the period is increased to 400.

6 Results

In order to provide some justification for the agent structure we have adopted we tested our problem against three agent architectures. The first, C_1, uses a single LCS with the sole objective of maximizing profit. It is like XCS except for the modifications described above and starts with a randomly generated rule set of 600 rules. The second, C_2, uses two LCSs, with 200 and 400 rules respectively, to meet the two objectives. Both of these are based on XCS and both start with randomly generated rules. The third agent structure is that shown in Figure 1 and described above. CS_A starts with 200 rules, and the rule set of CS_B is generated from the GCL up to a maximum of 400 rules. To test the systems we probabilistically generated 10 sets of 20000 environments, and ran each of the agents on these rule sets. To measure performance we kept a 365 day sum of profit received and a 365 day sum of maximum profit achievable. We refer to the ratio of these values, i.e. the moving annual percentage of attainable profit, as r. At the end of a run we freeze all adaptive mechanisms and we calculate the expected profit and the expected maximum profit. To do this we force the agent to exploit, then run through every possible environment generating an action, summing the probability of that environment occurring (since they do not occur with the same probability) multiplied by the profit achieved. We then divide this by the expected maximum to get a figure we denote $E(r)$. Table 1 shows

Table 1. Results for the three agent models averaged over ten runs of 20000 days

Model	Final r	Best r	$E(r)$
C_1	82.28 %	88.16 %	84.43 %
C_2	86.05 %	91.49 %	88.28 %
C_3	94.06 %	95.79 %	92.65 %

two things: Firstly the fact that both C_2 and C_3 outperformed C_1 shows that splitting the problem in two and using two LCSs (with the same total number of rules) to attempt to optimize the separate objectives improves the quality of the resulting strategy. Secondly, forcing the second LCS to concentrate on certain areas of the search space can also improve performance. To illustrate the second point and show some of the rules generated by C_3, Table 2 shows the optimum strategies for bidding when the unit is constrained on (a particularly profitable area of the environment space which occurs approximately one in twelve days). Almost all actions result in a profit when a unit is constrained on. When the unit is constrained on the environmental message matches 01********. The environment was constructed so that this area would reflect the real market. When a unit is constrained on it is forced to run and paid at its bid price. However, if it bids too high the scheduler will, if it can, choose another unit to constrain on. Bidding when constrained on can be summarised as follows: When capacity premium is maximum (********11), the best strategy is to bid as high as possible and still be in the unconstrained schedule for the times of peak demand (thus getting the capacity premium for those times). At other times the important factor in constraint bidding is the constraint levels (bits 5 and 6 in the environmental message). At the highest levels the best action is to bid the maximum (31). At level 1 the best action is 29 and at the lowest level the best action is 22 or 25, depending on the other variables. To give an idea of the rules CS_B uses, Table 3 shows some of the rules matching 01******

Table 2. A Summary of Optimum Actions for Bidding When Constrained On

Condition	Action	Meaning
01******11	8,9 or 20	CP at max
01**00**0*	22,25	Const. level 0,
01**00***0		CP not at max
01**01**0*	29	Const. level 1
01**01***0		
01**1****0	31	Const. level 2 or 3
01**1***0*		

exactly. We have not included the partially matching rules and rules under the MAM threshold for clarity. The first three rules have obviously been created

Table 3. A Summary of Optimum Actions for Bidding When Constrained On

Condition	Action	Prediction	Error	Condition	Action	Prediction	Error
01110*1111	0	712.94	0.53	01**0*0000	25	430.36	0
01**0*111*	4	485	0.4	011*0111**	20	367.94	0.04
011*00*11*	12	300.12	0.64	011**001*0	21	364.06	0.04311
011***1*0*	21	360.73	0.02	011**0000*	22	380.63	0.00053
01******00	22	365.27	0	01*0*1110*	28	492.66	0
011******0	21	377.89	0.03	01*001*10*	29	496.66	0.0012
010******0	22	380.72	0	01*1*1*10*	29	496.66	0
01**0*****	23	397.47	0	01*001***0	29	496.66	0
01*1*0****	23	397.22	0	01**100*00	29	496.66	0
01*1*0****	24	413.79	0.0003	01*001110*	29	496.66	0
0111000***	23	397.21	0.00045	011*11**00	29	463.51	0.00034
01**0*0010	25	430.36	0	010*1***00	30	502.09	0.0003
01**0*000*	25	430.36	0	011*11**00	31	529.82	0

from strong rules in other match sets (the reward levels for maximum capacity premium are higher). Whilst they may cause difficulty for a while their high error should result in their deletion or specialization in time. With the other rules a partially complete default hierarchy has emerged. There is a reasonable spread of more general rules for actions 20-25, with more specific rules with actions 28-31 for the environments with a higher constraint level.

7   Conclusions and Future Directions

The long term goal of this project is to develop a multi-agent adaptive model of the market, where each agent has to evolve strategies to compete and/or cooperate with eachother. Before we can examine the behaviour of agents in changing environments, it is important to know that the learning mechanisms of the agent are able to handle a static environment space. In this paper we have presented a static environment to an adaptive agent and given that agent a hierarchical structure using two LCSs tailored to meet different objectives. This model is not only is better at generating actions to maximize profit but also gives us a structure that allows the modeling of different objectives within a single agent. This is of particular importance in such a highly regulated market such as the UK Electricity Market, where real bidding is driven by other factors in addition to the simple maximizing of profit, such as maintaining market share and avoiding censure and/or fines from the regulator. We would also like to develop techniques applicable to a broader range of problems. Many real world decision making processes are similar to the model we have developed, in that they involve multiple, possibly related, objectives and reasonably well defined external factors that will effect the success of actions in meeting those objectives. The maintenance of a telecommunications network, for example, may involve regular decisions that effect how well a company meets its multiple objectives (such as meeting network demand, maximizing profit, maximizing market share, avoiding regulator punishment), while operating within the physical constraints of the network and predictable patterns of telephone usage. Whilst most real world problems are too complicated to accurately simulate, the investigation of simplified models constructed with expert knowledge of the system can still yield insights. In addition to particular simplified simulations of decision making scenarios, further investigation into modified reinforcement learning problems could aid our understanding of how competing requirements effect behaviour. For example, an Animat problem could be formulated where the food squares have quantities of food, and the agent could have dual objectives of making sure it can find food quickly and of maximizing its food intake over a longer period of time.

In addition to illustrating the possible advantage of using two LCSs, our experiments have shown that exploiting previously recognized properties of different classifier rule discovery mechanisms can improve performance, and that using limited size long term memory storage external to the classifiers can speed up convergence and help direct the search into profitable areas.

The next step in this project will be to start examining how several adaptive agents interact together in our simplified market, and to explore whether the alteration of objective weighting can produce explainable behaviour.

Acknowledgements. We would like to thank the National Grid Company for their support and advice on this EPSRC CASE sponsored project.

References

1. W. Fairney. Power generation in the 1990s. *Power Engineering Journal*, pages 239-246, December 1993.
2. F. Zhuang and F. D. Galiana. Towards a more rigorous and pratical unit commitment by lagrangian relaxation. *IEEE Transactions on Power Systems*,3(2):763-773, 1988.

3. M. Dorigo and U. Schnepf. Genetics-based machine learning and behaviour based robotics: a new synthesis. *IEEE Transactions on Systems*, Man and *Cybernetics*, SMC-23(1), 1993.
4. S. W. Wilson. Classifier fitness based on accuracy. *Evolutionary Computation*, 3(2), 1995.
5. M. Dorigo. Genetic and non genetic operators in ALECSYS. *Evolutionary Computation*, 1(2), 1993.

Evolution of Cooperation within a Behavior-Based Perspective: Confronting Nature and Animats*

Samuel Delepoulle[1,2], Philippe Preux[2], and Jean-Claude Darcheville[1]

[1] UPRES-EA 1059:
Unité de Recherche sur l'Evolution du Comportement et des Apprentissages,
Université de Lille 3,
B.P. 149, 59653 Villeneuve d'Ascq Cedex, France.
delepoulle@univ-lille3.fr,
darcheville@univ-lille3.fr
[2] Laboratoire d'Informatique du Littoral,
B.P. 719, 62228 Calais, France.
preux@lil.univ-littoral.fr

Abstract. We study the evolution of social behaviors within a behavioral framework. To this end, we define a "minimal social situation" that is experimented with both humans and simulations based on reinforcement learning algorithms. We analyse the dynamics of behaviors in this situation by way of operant conditioning. We show that the best reinforcement algorithm, based on Staddon-Zhang's equations, has a performance and a variety of behaviors that comes close to that of humans, and clearly outperforms the well-known Q-learning. Though we use here a rather simple, yet rich, situation, we argue that operant conditioning deserves much study in the realm of artificial life, being too often misunderstood, and confused with classical conditioning.

1 Motivations: Artificial Life and the Experimental Analysis of Behavior

As stated at its origins, artificial life deals with the study of life as it exists, and life as it might exist [18]. This paper deals with the first part of this project, the study of life as it exists. Our endeavor concerns the study of the dynamics of behavior relying on simple, though sound, nature grounded, assumptions. These assumptions are drawn from a selectionist approach of behaviors, compatible with, and complementary to, the selectionist approach of the evolution of living species. Basically, relying on Skinner's work, the selectionist approach of behaviors (or radical behaviorism, operant or instrumental conditioning [24,23,26]), states that a behavior is more likely to be re-executed, and spread in the population, if it is followed by positive consequences. Conversely, a behavior is likely

---

* This research is supported by "Conseil Regional Nord-Pas de Calais" (contract n 97 53 0283)

C. Fonlupt et al. (Eds.): AE'99, LNCS 1829, pp. 204–216, 2000.

to disappear if it is followed by negative consequences. Furthermore, behaviors being generally never redone twice exactly identically, variation naturally occurs in behaviors, hence a mutation-like mechanism is naturally included. This has much to share with the idea of genes being retained and spreading in a population if they are well adapted [25]. Before going any further, it should be clear that the selectionist approach is an alternative way of explaining behaviors (including humans behaviors) to that of the well-known cognitivist assumptions: brain act as a computer and process information. We do not wish to debate here about the validity of these approaches. We simply want to make clear our way of thinking. Operant learning is well known to be adequate to explain complex behaviors of animals, and to let animals acquire complex behavior [26]. These behaviors are indeed much more complex than those currently exhibited by top of the art robots [5]. In [7], we have shown that a task consisting in sharing a task between agents can be solved quasi-optimally by a set of agents which behavior evolution is selectionist. In this paper, we focus on the emergence of cooperation among a set of agents and explain it by means of a selection of behaviors due to the context in which they are emitted, or "contingencies of reinforcement". Our approach is original is many respects. First, we try to keep a pure behavior-based approach, relying on the work of the experimental analysis of behavior. Second, we work in the same time on living beings, mostly human beings, and computer simulations. Computer simulations compel us to define very precisely the principles of selection of behaviors and give us the opportunity to simulate them very strictly, showing their validity, and showing their weaknesses. This research might lead to new ideas on the issue of making cooperation between adaptive agents possible. In the world of agents, cooperation is considered as a main issue, and as a problem to solve in itself. Stemming from the analysis of behavior, we wish to put forward the point of view that cooperation can also be considered as a "natural" adaptation of agent's behavior to their environmental contingencies, as far as the agents are indeed able to adapt themselves.

2   Evolution of Cooperation

What makes the evolution of cooperation possible between two, or more, living beings? In natural situations, one living being frequently behaves in such a way that produces favorable consequences for other living beings. For the individual who emits such a behavior, the immediate payoff can be inexistent, or negative. Hence, the individual does not directly benefit from his, or her, own behaviors. How can such behaviors appear? How can they be retained? These are serious problems arising if we adopt a selectionist way of thinking, such as within the theory of natural selection, or the selection of behaviors by way of their consequences [25].

In his famous work [2], Axelrod uses the problem of the Iterated Prisoner's Dilemma (IPD) to formulate the problem of cooperation in the framework of game theory. He demonstrates the importance of the repetition of cooperation situations. Dawkins [6] studies the role of genes in cooperative situations. Biologists works show the importance of social behaviors [11,34] Many other attempts to explain cooperative behaviors have been made [16,12]. However, none of them have yet tackled the question of the evolution of cooperation within a strict behavioral perspective. In this paper, we study the conditions within which cooperation can appear and get installed.

We propose to examine a rather simple situation in which cooperation can appear. We do not work with the IPD because we think IPD is inadequate for our purpose with several respects. In Particular because agents act synchronously and are perfectly aware of consequences of their behaviors.

Different points of crucial importance have been let apart with regards to more realistic situations. In nature:

- the consequence of a behavior may be delayed so that the relationship between a behavior and its consequence gets blurred and eventually hard to figure out,
- lots of behaviors are not rewarded: they are neutral with regards to their consequences, hence their selection. Due to the parallelism between the selection of behaviors and the selection of genes, we emphasize this point because a crucial importance of "neutral mutation" in the evolution of living species is suspected.

Dealing with the study of the evolution of cooperation in natural/real-world situations, we design an experimental situation relaxing IPD limitations and involving the previous points. According to a straightforward Skinner's heritage, cooperation naturally emerges, if it ever does, thanks to an exchange of reinforcements between the agents[1].

We have worked along a line that confrontates the dynamics of behavior of human beings with that of animats merely implementing reinforcement algorithms. We put a fundamental emphasis on this confrontation for different reasons. To be clear, we obviously do not have the ambition to simulate a human being (!): the algorithms that we use are much cruder. More seriously, for the problem of cooperation in a certain experimental layout, we aim to compare the dynamics of human behaviors with that of reinforcement algorithms. Then, relying on the fact that the algorithms are all selectionist, we suggest that human beings may behave in the same situation according to selectionist principles if both dynamics match. Furthermore, this would suggest that a social behavior can be the consequence of individual behaviors. The aim is also to assess the performance of reinforcement algorithms in environments in which we try to carefully keep the very essential features of the real world.Based on insights taken from naturally evolved structures, these algorithms are seldomly put into environments that somehow mimics the environment they evolved in and to which they have adapted. We think that some "weird"or deceiving performance of these algorithms when used in artificial environments (related to combinatorial optimization for

---

[1] it is remarkable that cooperation emerges at a rather identical rate and pace whether the human subjects have been completely told the situation before the experiment, or not, the latter being the normal way the experiment has to be conducted.

instance) is simply due to the fact that they are suited to the complexity of the environment in which they evolved and not to the artificial environments in which we sometimes use them.

In the following, we first develop on the idea of cooperation and we describe the experimental situation we have designed. Then, we briefly describe the reinforcement algorithms that were used, most of them being directly inspired by principles of the analysis of behavior. Based on simulations, we discuss their ability to evolve cooperation. Finally, we draw some conclusions and discuss various perspectives of this work.

3   A Behavioral Perspective on Human Cooperation

3.1   Description of a "Minimal Social Situation"

As stated by Hake and Vuklich [10] within a behavioral framework, cooperation is defined by the fact that the reinforcement of two individuals must be *"at least in part dependent upon the responses of the other individual"*. In cooperative procedures, an individual can improve not only his own payoff but also the payoff of his, or her, party.

In order to understand the development of cooperation, we use a very simple situation, namely a "minimal social situation". In that situation, subjects (humans) can interact through a controlled device, a computer. Each response of a subject is made on a computer, each subject being located in a separate room, and being unaware of the presence of his, or her, party. Thus, they are not told that they are in a social situation. As a matter of fact, most subjects claim at the end of the experimentation that they were alone in front of a computer, without any interaction of any kind with anything (*e.g.* an animat), or anyone else.

The experimental layout is directly inspired by Sidowski's experiments [21, 22]. In this situation, subjects are invited to interact with a computer during thirty minutes. On the computer screen, a window includes two buttons and a counter. When subject A (resp. B) clicks on button 1, B's counter (resp. A's) is incremented (rewarding action, or R action). When A (resp. B) clicks on button 2, B's counter (resp. A's) is decremented (punishing action, or P action). Then, the behavior of a subject has no result on his, or her, own counter but on the counter of his, or her, party. It is noteworthy that a subject may choose to do nothing for a while, or even during the whole experiment. Furthermore, agent actions are not synchronous: they do not have to click at the same time and one click is immediately taken into account to provide its consequence to the other subject. This is a fundamental difference between the situation we use and other published work in the field.

3.2   Results

Thirteen couples of subjects underwent this experiment. During the first minutes of the experiment, clicks are equally shared on the two buttons but after a few minutes the rate of clicks on the button 1 increases. At the end of the experiment, subjects click about five times more on button 1. These results accord with other studies using similar procedures with humans [21,22], and with animals [4].

3.3   Interpretation

According to the experimental analysis of behavior [24,26] we can suggest how cooperative behaviors emerge. In this approach, behaviors are selected by their consequences [9]. The behavior of an organism varies; if these variations are profitable to this organism, they are likely to be selected.

In the experiment we made, if the action (R, or P) of both subjects are done synchronously, three combinations are possible.

- If both subjects press button 1, each one gains 1 point. In that case, clicking on button 1 is reinforced and they will continue clicking on button 1,
- If both subjects press button 2, each one loses 1 point. So, they will change their behavior and press button 1. Then, the subjects comes to the previous situation,
- If one subject presses button 1 while the other one presses button 2, the first subject loses 1 point while the second gains 1 point. The first will change his, or her, behavior and the second will not. Then, subjects are in the second situation, which again leads to the first.

Table 1. This table gives the payoff for both subjects of their actions. The first payoff (±1) of the couple is that of A, the second is that of B. See the text for explanations of the arrows.

B\A	clicks on button 1		clicks on button 2
clicks on button 1	+1 +1		+1 -1
		↖	↓
clicks on button 2	-1 +1	→	-1 -1

Table 1 displays the consequences of the subject's choices if they act synchronously. If their actions are synchronous, the spaces of states is made of 4 states. Arrows show the trajectory among the states. State ($+1$ $+1$) is an attractor for all initial states. Thus, cooperation is an attractor for the dynamics of behaviors in this experimental situation: if agents behaves synchronously, cooperation appears very quickly and remains. We observe this phenomenon experimentally: originally behaving asynchronously, cooperation appears in the group exactly at the very moment when responses become synchronized.

Each pair of subjects can be represented in a two dimensional space, the axis being the cooperative rate of the two subjects. Figures ?? and ?? represent the evolution of the cooperative rate of the 13 couples of subjects. In figure ?? representing the begining of the experiment, we notice that reponses form a two dimensional gaussian-like distribution at the center. That distribution may be the result of random responses of the two subjects. Figure ?? shows the result during the four last minutes of the experiment. The major part of the distribution is concentrated on a corner, this point represents the maximum cooperation rate for both subjects. Considering the results of each group, we notice that when a pair of subjects is entering in that state, they always stay in it. After a certain amount of time (very variable), most of groups switch to that state.

To finish the analysis of the situation, we notice that a mere stimulus-response architecture is not able to solve the minimal social situation even in this very restricted case where each agent can either reward or punish his, or her, party, and actions are synchronous. A response-stimulus architecture is required which opens up the road to operant conditioning.

4   Reinforcement Learning Algorithm Facing Cooperation

The interest of the experiment is not to show that human subjects can learn to behave in very simple social situations because we know that humans can learn much more complex social situations like imitation or verbal behavior. The interest is to show that a social situation can be explained by the knowledge of the individual law of behavior. Social organization can be the emergent result of individual behaviors. This analysis can also be supported by experimental work on social insects. It was shown that insects living in societies (ants for example) can perform complex tasks without the presence of an individualized central organizer [8,30]. Many works have used this model to construct self-organised population of agents (see [19,20] for instance).

The previous section has shown how the behavior of humans subjects can be explained using the principle of behavioral selection. So we propose to use agents based on a selectionist architecture. Reinforcement learning algorithms (RL) meet this requirement [17].

We have used five reinforcement-learning architectures, four of them being directly originated from the analysis of behavior while the fifth is the well-known Q-learning algorithm.

An agent can emit only three behaviors. It can reinforce its party (R), punish its party (P), or do nothing (N). We introduce the "do nothing" behavior to have an asynchronized situation at the beginning of the simulation. As we have shown in the experimental section, synchronization makes selection of cooperation easier because the consequence of a behavior is received immediately.

In the sequel of this section, we present the 5 reinforcement learning algorithms very briefly and give the experimental results in the next section.

4.1   The Law of Effect

At the end of the XIXth century, Thorndike [31,32] has studied the animal behavior and he has suggested a law to predict the evolution of behaviors with

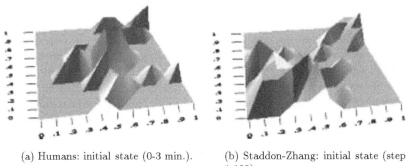

(a) Humans: initial state (0-3 min.).   (b) Staddon-Zhang: initial state (step 1-100).

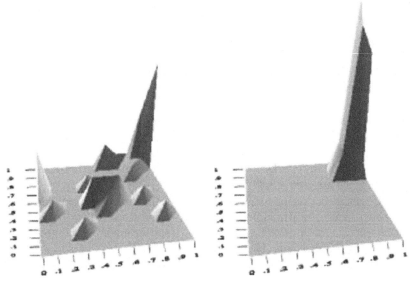

(c) Humans: final state (26-29 min.).   (d) Staddon-Zhang: final state (step 901-1000).

Fig. 1. Graphical representation of the cooperative rate of the couples of subjects (1st column) and Staddon-Zhang simulation (2nd column) at the beginning of an experiment (a) and (b), and at the end of the experiment (c) and (d). The abscissa is the rate of clicks on button 1 of subject, or agent, A, the ordinate is the rate of clicks on button 1 of subject, or agent, B. At the beginning of the experiment, the distribution is rather gaussian at the center, meaning that clicks are somehow equally shared between the two buttons. At the end of the experiment, there is a clear spike corresponding to the cooperation: both subjects, or agents, click mostly on button 1. Of course, some noise always occurs which explains the little blobs here and there.

regards to their consequences (positive or negative). We formalize his "law of effect" as follows:

$$\text{if } C_i \neq 0, \text{let } s = \frac{C_i.\alpha}{|C_i|}, \text{then} \begin{cases} p_i = \frac{p_i+s}{1+s} \\ p_j = \frac{p_j}{1+s} \text{ for } j \neq i \end{cases}$$

where p_i is the probability of apparition of a behavior i, C_i, the consequence of behavior i and α is the learning rate, that is the relative weight of current stimuli with regards to behaviors emitted previously.

4.2   Hilgard and Bower's Law

In Hilgard and Bower's law [13], very similar to the law of effect, named "linear reward–inaction algorithm", all non-reinforced actions are weakened. This algorithm always converges with a propability 1 on a particular action (but not always on the best action). This can be expressed by:

$$\text{if } C_i > 0 \begin{cases} p_i = p_i + \alpha(1 - p_i) \\ p_j = p_j - \alpha p_j \qquad \text{for } j \neq i \end{cases}$$

with p_i, C_i and α defined as in the law of effect.

4.3   Staddon-Zhang Model

In 1991, Staddon and Zhang [27] proposed a model in order to solve the *assignment-of-credit* problem without teacher (unsupervised learning). Staddon and Zhang show that this model accounts for qualitative properties of response selection. Their model accords itself not only with regular data in behavior analysis but also with "abnormal" behaviors (autoshaping, superstition and instinctive drift).

In this model, each behavior has a value V_i. All values V_i are in competition and a "winner-take-all" rule is used to select the behavior to emit at each time slot. The values V_i are updated at each time slot according to the equation:

$$V_i = \alpha V_i + \epsilon(1 - \alpha) + \beta V_i$$

where $0 < \alpha < 1$ (α is a kind of short term memory parameter), β, the reinforcement parameter should be positive for rewards and negative for punishments, and ϵ is a white noise.

4.4   Action-Value Method

The goal of this method is to estimate the mean consequence for each behavior and to choose the best one to emit in order to optimize the reward. Sutton [29] gives an iterative method to calculate this estimation. V_i is the estimation of mean consequence and N_i is the number of occurences of behavior i in the past. If behavior i is emitted, then

$$\begin{cases} V_{i+1} = \frac{1}{N_i}[C_i + (N_i - 1).V_i] \\ N_{i+1} = N_i + 1 \end{cases}$$

Such a method converges quickly if we allow it to explore different behaviors. So in complement to the behavior driven by the previous equations, the agent is allowed to emit a random behavior with probability ϵ.

4.5   Q-Learning

Q-Learning is one of the most famous reinforcement learning method. Based on Sutton and Barto's Time Derivative (TD) model [28], this method has been proposed by Watkins [33]. Q-Learning is an algorithm for solving rapidly and easily stochastic optimal control problem. $Q_{s,a}$ represents the expected future payoff for action a in state s. Q-Learning works by modifing $Q_{s,a}$ for each pair of state-action using the following equation:

$$Q_{s,a} = Q_{s,a} + \alpha[r + \gamma \max_b Q_{s',a} - Q_{s,a}]$$

In a markovian environment and if the problem is stationary, it was shown that this algorithm converges with probability one to the optimal value. In practice, Q-Learning does not explore the environment sufficiently. Thus, it has been suggested to produce variability by adding noise. Contrary to the four other reinforcement learning methods presented above, Q-Learning is not "context-free": if a behavior is reinforced only in a precise situation, the animat will be able to emit this behavior in that situation only.

5   Results of Simulation

Agents are tested in two situations. For the first(individual situation), they can choose between three behaviors which have direct consequences on their own counter. In the second situation (interactive situation), agents are put by couples in the same condition than human subjects. The algorithm performs a number of behavior that is rather similar as for human subject

In the individual situation (see columns 2, 3, and 4 of table 2), the environment is very simple and all algorithms succeed in adapting their behavior to optimize their rewards. The difference between algorithms is the speed of convergence. The only algorithm which exhibits a different behavior is Staddon-Zhang's law: within about thirty percent cases, it has a behavior which yields no consequence at all, which may seem a major weakness. However, a careful examination of the behavior of this algorithm shows that it keeps exploring its environment while all other algorithms have settled into a rest point.

Table 2. Distribution of different behaviors among a hundred agents after one thousand steps of simulation. R represents the behavior that provides reinforcements to its party, P is the bahavior that provides punishment and N is doing nothing. Reinforcements can be given to itself (individual situation) or to the other agent (social situation).

algorithm	individual			social		
	R	P	N	R	P	N
law of effect	100	0	0	53	16	31
Hilgard and Bower	100	0	0	35	28	37
Mean earning	89	7	4	54	42	4
Q-Learning	89	7	4	58	17	25
Staddon-Zhang	66	0	34	86	1	13

In the "social" situation (see columns 5, 6, and 7 of table 2), results are opposite. By far, results of Staddon-Zhang are the best: at the end, it cooperates nearly nine times out of ten. Only Staddon-Zhang, the law of effect, and Q-Learning exhibit results that differs significantly from a random behavior. The very good results of the Staddon-Zhang's algorithm clearly show that the performance of a method designed for stationary problems can be very different in a dynamical situation. Furthermore, as said before, amongst the 5 algorithms, the algorithm relying on Staddon-Zhang's law has the unique feature to keep exploring its environment. This feature might be held as a weakness in a stationary environment, but it is the source of its strength in a dynamical situation. Cooperative rate are plotted on figure **??** and **??** at the beginning and at the end of simulation from random behavior, the algorithm comes to mostly cooperate.

One can be suprised of the very good results of Staddon-Zhang's model with regard to Q-learning known as the best reinforcement learning algorithm. We must take care that Q-Learning is based on the prediction of the amount of reinforcement but in the minimal situation, prediction of reinforcement is not relevant. As shown in table 1, for this extreme situation the score of one agent is not dependant on his own behavior but on it's party's behavior. The prediction of the amount of reinforcement in function of the behavior does not produce a significant result, which is also the reason why the algorithm "mean earning" is not really good in that situation.

In contrast, Staddon-Zhang is a method based on optimization of the behavior himself regarding his consequences. This might be the method used by humans during the game. Staddon-Zhang can take advantage of the situation and reward the other agent when it addresses reinforcement because it is directly inspired of the evolution of animal behavior, even in abnormal situation [27].

6   Discussion and Perspectives

By using a minimal social situation, we show how cooperation, is possible if the behavior of subjects is selected by their environment. The interest of such situations is twofolds. First, there can be a precise recording of emitted behaviors so that analysis is possible. Second, they can be simulated by simple adaptive agents.

The minimal social situation has shown that the asynchronization of agents' behavior is an important issue since synchronization implies cooperation straightforwardly with a high probability. So, the emergence of the synchronization of the agents' behavior is an important part of the route towards cooperation. More generally, it should be clear that the synchronization of actions, which generally lies implicitly in the background of lots of works dealing with simulations, is not a mere detail.

By using a *minimal social situation*, we may suppose that behavioral selection can be important to account for the emergence of cooperation. The use of adaptive agents in the same situation supports such an analysis. So, adaptive agents are a precious tool to test hypothesis about social behaviors. However, we emphasize that simulations never proove anything firmly. Reciprocally, experiments based on human or animal behavior provide insights into the design of agents. In many cases, animal behavior, as the result of million years of evolution, is able to optimize very complex situations. By knowing the mathematical relation between physical characteristics of environment and behaviors, we might be able to build a new generation of architectures of artificial agents, able to learn in many situation. This kind of agents should be able, for instance, to learn to imitate and to learn from verbal instruction [14,15].

This procedure — controlled social situation and simulation by agents — will be used in order to study more complex behaviors such as work division or sequential decision making in social situations. If the situation is more complex, many reinforcement learning algorithms used in this paper may become unusable because, apart for Q-Learning, they are "context free". Hence their behavior cannot be truly controlled by the characteristics of their environment, unless they are adapted to integrate the context. Due to its performance in a dynamical environment as well as its soundness with regards to the experimental analysis of behavior, we will work towards making Staddon-Zhang's law context sensitive.

References

1. Attonaty, J.M., Chatelin, M.H., Garcia, F., and Ndiaye S.M., Using extended machine learning and simulation technics to design crop management stategies. *In EFITA First European Conference for Information Technology in Agriculture*, (1997) Copenhagen
2. Axelrod, R., *The evolution of cooperation*, Basic Book Inc. (1984)
3. Bergen, D.E., Hahn, J.K., Bock, P., An adaptive approch for reactive actor design, *Proc. European Conference on Artificial Life*, (1997).
4. Boren, J.J., An experimental social relation between two monkeys, *Journal of the experimental analysis of behavior*, **9** (1966) 691–700.
5. David S. Touretzky, Lisa M. Saksida, Skinnerbots, *Proc. 4th Int'l Conf. on Simulation of Adaptive Behavior, From Animals to Animats 4*, Maes, Mataric, Meyer, Pollack, Wilson (eds), MIT Press, 1996
6. Dawkins, R., *The Selfish Gene*, Oxford University Press, Oxford (1976)

7. Delepoulle S., Preux Ph., and Darcheville J.C., Partage des tâches et apprentissage par renforcement, *Proc. Journées Francophones d'Apprentissage*, (1998), 201–204 (in french)
8. Deneubourg, J.L., and Goss S., Collective paterns and decision-making, *Ethology Ecology and Evolution*, **1** (1989) 295–311.
9. Donahoe, J.W., Burgos, J.E., and Palmer, D.C., A selectionist approach to reinforcement, *Journal of the experimental analysis of behavior*, **60** (1993) 17–40.
10. Hake, D.F., Vukelich, R., Analysis of the control exerted by a complex cooperation procedure, *Journal of the experimental analysis of behavior*, **19** (1973) 3–16.
11. Hamilton, W. D., The Genetical Evolution of Social Behaviour, *Journal of Theoritical Biology* **7** (1964) 1–52.
12. Hemelrijk, C.K., Cooperation without genes, games or cognition, *Proc. European Conference on Artificial Life*, (1997).
13. Hilgard, E.R. and Bower, G.H.: *Theories of learning* (fourth edition). Prentice-Hall, Enblewood Cliffs, NJ.
14. Hutchinson, W.R. Teaching an agent to speak and listen with understanding: Why and how? Proceedings of the Intelligent Information Agents Workshop, CIKM, Baltimore: http://www.cs.umbc.edu/ cikm/iia/submitted/viewing/whutchi.html
15. Hutchinson, W.R., *The 7G operant behavior toolkit: Software and documentation*, Boulder, CO: Behavior System.
16. Ito, A, How do selfish agents learn to cooperate? *Proc. Artificial life 5*, Langton, Shimohara (eds), MIT Press, (1996) 185–192.
17. Kaelbling, L.P., Littman, M.L., and Moore, A.W., Reinforcement learning: a survey, *Journal of Artificial Intelligence Research*, **4** (1996) 237–285
18. Langton Ch., Artificial life, *Proc. Artificial Life*, Langton (ed), Addison-Wesley, (1987), 1–47
19. McFarland D., Towards Robot Cooperation, *Proc of the International Conference on Simulation of Adaptive Behavior: From Animals to Animats 3*, (1994) 440–444.
20. Murciano, A., Millán, J.R., Learning signaling behaviors and specialization in cooperative agents. *Adaptive Behavior*, **5(1)** (1997) 5–28
21. Sidowski, J.B., Reward and Punishment in a Minimal Social Situation, *Journal of Experimental Psychology*, **55** (1957) 318–326.
22. Sidowski, J. B., Wyckoff, B., and Tabory, L., The influence of reinforcement and punishment in a minimal socila situation. *Journal of Abnormal Social Psychology*, **52** (1956) 115–119.
23. Skinner, B.F., *The behavior of organisms*, (1938). Englewood Cliffs, NJ: Prentice Hall.
24. Skinner, B.F., *Science and human behavior*, (1953) New York: Macmillan.
25. Skinner B.F., Selection by consequence, *Science*, **213**, 501–514, 1981
26. Staddon, J.E.R., *Adaptive Behavior and Learning*, (1981) Cambridge University Press.
27. Staddon, J.E.R., and Zhang, On the Assignment-of-Credit Problem in Operant Learning, *Neural Network model of Conditioning and Action*, M.L. Caumais S. Grossberg (eds), (1991) Laurence Erlbaum : Hillsdale, N V.
28. Sutton, R.S. and Barto, A.G., "Time-Derivative Models of Pavlovian Reinforcement", *Learning and Computational Neuroscience: Foundations of Adaptive Networks*, M.Gabriel and J.Moore (eds.), (1990) 497–537. MIT Press: ftp://ftp.cs.umass.edu/pub/anw/pub/sutton/sutton-barto90.ps
29. Sutton, R.S., *Reinforcement Learning*, MIT Press (1998).
30. Theraulaz G., Pratte M., and Gervet, J., Behavioural profiles in *Polistes dominulus* (Christ) wasp societies: a quantitative study. *Behaviour* **113**, (1990) 223–250.

31. Thorndike, E.L., Animal Intelligence: An experimental study of the associative process in animals, *Psychology Monographs*, **2** (1898)
32. Thorndike, E.L., *Animal Intelligence: Experimental studies.* (1911) New York : MacMillan.
33. Watkins, C.J.C.H. and Dayan, P., Q-Learning. Technical Note. *Machine Learning*, **8(3)**, (1992), 279–292
34. Wilson, E.O., *The Insect Societies.* (1971) The Belknap Press, Harvard University Press.

Evolving Behavioural Animation Systems

David Griffiths[1] and Anargyros Sarafopoulous[2]

National Centre for Computer Animation, School of Media Arts and Communication,
Bournemouth University, Talbot Campus, Fern Barrow, Poole, Dorset BH12 5BB,
United Kingdom.
dave@blueammonite.f9.co.uk `Asarafop@bournemouth.ac.uk`
`http://www.blueammonite.f9.co.uk/alifeart`

Abstract. One of the features of Artificial Life (AL), is its ability to
cross boundaries between traditionally separate disciplines. While its foundations are in biological research and computing, its visual nature has
implications for the fields of art and entertainment. Using a continuous
genetic algorithm, adaptive autonomous agents can explore user created environments. If these agents have pressures of 'natural' selection
imposed on them, they can exploit the environment and create simple
solutions to survive. When the environment becomes complex enough,
the emergent solutions can, in turn gain in complexity, leading to un-
expected and visually pleasing results. We produce animation sequences
whose content/aesthetic is defined by the foraging and mating behaviour
of simulated agent colonies.

1 Introduction

Artificial Life systems represent a promising method of creating visually interesting and organic animation. A wide variety of AL systems have been created
for many uses. The first examples of agent based AL were purely abstract, in-
habiting one-dimensional environments (Tom Ray [5]). As these were difficult to
visualise, they were later expanded into two-dimensional forms. Sims [6] concentrates more on genetically generating individual movement and behaviour in a
physically based world, while Yeager [9] creates ecologically based environments
to contain populations of autonomous agents. The visual content of such systems are arguably their most important component, as they are the only way
to understand the dynamics of their behaviour, and draw comparisons with their
natural world counterparts. To date, systems written to fully explore the inhe-
rently visual properties of these techniques have been restricted to the evolution
of texture [7] and shape, Todd and Latham [8].

The artificial life simulation detailed here, is designed to be used as a tool
for exploring the visual properties of emergent behaviour. The focus of this
simulation are the creatures which inhabit it. The environment is where all the
behavioural elements are combined, and where the creature's survivability is
put to the test. This survivability is the driving force behind the evolutionary
algorithm, which is based on a GA with fixed length integer representation. The

C. Fonlupt et al. (Eds.): AE'99, LNCS 1829, pp. 217–227, 2000.

creature's behaviour is encoded in an artificial neural network and the resulting animation is a visual expression of this data, through the movement and extra detail via the shaders.

2   The Environment

The environment entirely defines the behaviours of the creatures, which evolve within it. The environments used here consist of several simple elements for the creatures to interact with. (see fig. 1) These elements are designed to present problems which are an broad abstractions of those found in the real world. It is hoped that this strategy will encourage actions in the creatures which are recognisable by us. The elements in the environment are food and barriers, creatures are free to roam in this environment in three dimensions, but may not intersect each other. Food is represented by small spherical objects, their distribution is interactively defined. Barriers are rectangular objects, which impose a small penalty on the creature's energy supplies if they collide. Barriers may be used to contain the creatures in a finite space, or to break up the environment in a more complex way.

Fig. 1. An Example Environment.

Colour is an extremely important part of the environment. In fig 1, barriers are shown as semi-transparent blocks, the small objects are food and the other multicoloured objects are creatures. All objects have a colour assigned to them, it is the only way the creatures can differentiate between different objects.

A creature's handling of the energy flow of the environment largely dictates it's implicit fitness. All energy in the environment comes from the food (apart from that contained in the first generation of creatures) so to obtain energy to live and survive; a species has to learn to eat.

The other source of energy available to the creatures is obtainable through fighting. If a creature activates its fighting action, it may be able to take a 'bite' from its prey's energy store. This way, energy can flow from creature to creature. This method of predation is implemented to encourage specialisation into predator and prey.

The maximum energy that a creature can store is set by it's size, the larger the creature, the more energy it can carry.

There are two modes of energy use in an environment:

1. "Energy loss environment", where energy is constantly being used up by creatures - and has to be constantly replaced. This is the most physically accurate.
2. "Circular energy environment", where there is no energy loss. All energy used by a creature is deposited in food when and where it dies. This approach seems to encourage very lazy creatures which live together in densely packed groups, the reason being, that they don't ever have to move very far to get energy.

Energy is used up in most of the creature's actions. Firstly, there is an energy price taken off all the creatures every frame. This price is proportional to the creature's size. Energy is also taken off when a creature moves, relative to the speed of the movement. When a creature gives birth to its children, an amount of the parent's energy is given to each of the offspring (this amount is defined in the genome), if this value is too high, the drain of energy will kill the parent. This should not be seen necessarily as a disadvantage; the genes of the parent will have been passed on. If this value if too low, the young creatures will have very little time to find food before they die.

3   Agent Architecture

The visual architecture of a creature is used as a method of displaying its individual attributes. There are three main attributes, which dictate a creature's abilities. Action distance, or reach, Sense distance, or how far it can see, and maximum speed. These attributes are defined in its genes by a ratio to encourage specialisation, and are expressed by scaling the creature's three body pods. The three representational pods are shown in fig. 2, and represent the following:

Clockwise from top right on the creature are the spiky action pod, smooth sense pod and rippled motion pod. Holes in the surface indicate energy loss. The effects of selection are clearly visible through the shaders, which change the surface appearance, and also the final geometry of the creatures. These shaders are part of the renderer.

Aggression is expressed through the use of a spiky displacement shader on the creature's action pod. Movement expressed through a rippling displacement shader on the creature's transport pod, which simply projects a sine wave through the object. The speed of the rippling is modulated by the current speed, and the wavelength is also modulated by an arbitrary gene connection.

Fig. 2. Creature showing the shaders used to indicate it's attributes.

The Creature's surface shader was the most complex shader in the system (see section 6). It incorporates the age, energy level and colour of the creature. The age of a creature is visualised by interpolating between different shading functions. With a young creature the shader is even, as it ages, the shader becomes distorted. Using "toon shader" techniques, the shader also gradually loses its smoothness, and eventually comprises three colours. A transparency component is used to show the energy level. As a creature loses it's energy, semi transparent holes appear on the creature, which grow until either energy is replenished, or the creature becomes completely transparent, and dies from lack of energy. To add to the creature's pictorial individuality, various arbitrary genes map to shader attributes, which control colour distortions on the surface of the creature. This serves the purpose of increasing the visual indications of mutations and variability.

3.1 Behavioural Modeling

The only information a creature has of its environment is the colour and relative position of the nearest object. This means that to be successful they must associate certain objects with certain colours, and actions.

There are 4 basic building blocks of the creature's behaviour; some of them are necessary for the creature to learn before it can survive on its own, others are more optional behaviour.

1. Eating
   This is the first thing that a species has to learn, as it must gain energy to move about its world.
2. Mating
   If a species is to continue to adapt and evolve then mating is a necessary action to learn. It is also beneficial (or though not vital) to learn to recognize members of it's own species.

3. Fighting

Fighting is not necessary for survival, although it is a secondary resource of energy. Some species practice cannibalism to some extent.

4. Signaling

Signaling does not have any immediate benefit for a creature, but it can lead to some of the most interesting results. To signal, a creature changes it's surface colour to that of the signal colour stored in its genome.

3.2 Artificial Neural Networks

As in existing artificial life systems [9], Artificial Neural Networks (ANN's) are used to control the creature's behaviour. The ANNs implemented here, are versions of feed forward and recurrent networks. The difference between other such implementations is the lack of a learning function. When a creature is born, its ANN is hardwired, and cannot be changed. Despite this limitation, the GA is able to select creatures with successful modifications in their ANN topology.

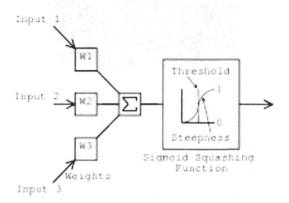

Fig. 3. Neuron Diagram.

These neurons are networked together as described in section 3.2.2 to form ANN's with eight inputs and eight outputs and seventeen neurons. The amount of neurons was fixed, due to the nature of the genome representation. The small amount of neurons helped with the speed of the simulation. The ANN inputs simulate the creature's perception of the world, and are:

1. Relative Angle of nearest object.
2. Colour of nearest object.
3. Current energy level.
4. Current age.

The creature is also given information on its internal status, how much energy it has left (or how hungry it is) and it's current age. Four outputs control the direction and speed of the creature, and four outputs are dedicated to the action selection of the creature, which simply uses the highest activated output to select the action (Eat, Mate, Fight, Signal).

Fig. 4 shows an example ANN taken from the net view window, inputs are at the top, outputs at the bottom. Activated neurons are indicated by their brightness. To add a little realism to this simple structure, a small amount of noise was added to each of the inputs

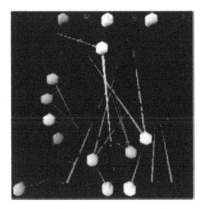

Fig. 4. Neural net visualisation.

3.3   ANN Genome Encoding

The genome for each creature supports a description of seventeen neurons, each using three weighted analogue inputs. The threshold value and steepness of the sigmoid squashing function are contained within the genetic representation for each neuron.

The attributes section details other factors of the creature's description, such as signal colour, and number of children - these do not directly affect the ANN. Each ANN has eight inputs and outputs. The topology stored in the genome describes the source of each neuron input and ANN output. The neuron definitions were separated in this way to allow meaningful crossover, which would not be liable to break the networks. The ANN's seemed quite resilient to mutational changes in their structure.

3.4   ANN Topological Approaches

Two different methods were used to describe the architecture of the ANN, the first was a fairly rigid two layer ANN. The first eight neurons were connected

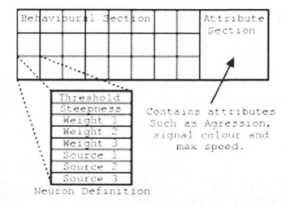

Fig. 5. Genome mapping, incuding the ANN description.

exclusively to the inputs of the ANN, while the second layer were permitted to connect to the first layer or the inputs (to allow for jump ahead connections). The outputs connected to the second layer. This structure allowed the GA to start with a fairly promising topology, and meant that creatures of early generations were generally relatively fit. However, this method was too inflexible to allow interesting structures to be found, and it also lacked the temporal features of a recurrent network.

The second method allowed a much more freeform architecture, including recurrent connections. Any neuron could simply connect to any ANN input or any neuron output. ANN Outputs could connect to any neuron output. This structure allows neurons to feed back into themselves via other neurons, which creates temporal properties, neurons may oscillate, or fire off each other in complex ways. This can also be seen as a very simple form of memory.

4   Evolutionary Algorithm

A species of creature becomes viable when it is capable of gathering enough energy to mate and reproduce its genetic code. Until this point, help is given via a GA style fitness function. Fitness is awarded incrementally for various activities: choosing any action, eating an item of food or successfully reproducing. If no viable creatures exist, the environment will eventually empty and the fittest creature can then be reloaded into a new generation with mutations.

The mating function activated by a viable creature incorporates crossover and mutation to create the child from the parent's genomes. Once this becomes the sole method of creature reproduction, the explicit fitness function becomes redundant, and the evolutionary method can be seen as modeling "natural" selection, i.e. an implicit selection based on reproduction.

5   Visual Representation: 3D Real-Time Display

The three dimensional environment is displayed in real-time for previewing creature's behaviour. The viewing camera may be moved around the environment interactively, or set to follow an individual creature for more detailed examination of behaviour. Extra information can also be shown, such as the sense and action distance for each creature. Graph, population viewing and family tree view windows are all part of the real-time system.

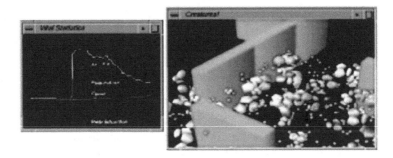

Fig. 6. Statistics window and Preview window.

6   Rendered Output

Final output for the animation is a high quality rendering. Each creature is represented by three primitive spheres and two hyperbolas, which connect them. The creatures are all shaded using their genetic textures. Barriers are shaded as semi transparent to allow creatures to be seen behind them. Fog can be used to represent their general viewing distance. The system communicates all the gene information to the renderer, and converts it into shader arguments.

7   Results

The initial purposes of this system were to generate lifelike animation, which had an interesting visual quality. However, the effectiveness of the solutions found can be examined in a more rigorous benchmark foraging test. This test consists of a random path of food, which provides an accurate idea of how effective the individual's foraging strategy is.

Another indication of effectiveness of a species is through the ecology of the environment. If the amount of food in the environment is low, while the amount of creatures or the reproduction level is high, then this either reflects favorably on the species, or tells us that the environment is too easy. Even in a situation where the environment is too easy, progress will still be made, as competition between the creatures will be high.

Fig. 7. An example species bred in an environment containing barriers.

Many species have been evolved; most of the interesting ones are described below. It was surprising that from such a simple system, so many solutions and traits emerged.

The first problem to be overcome was that of gaining energy, or feeding. Two broad strategies were employed, one of random path following, in the hope that food would be accidentally found at some time, and one of active food foraging. The creature that seemed to do this best (the "Blue forager") was tested in the benchmark environment. The mechanism it uses is to look around rapidly until an object is sighted. If this object has a high green colour channel then the creature will fire it's forward neuron, and activate its desire to eat. It is not known at what point in the generation this fairly simple strategy appeared, but in subsequent tests, it has occurred again. The effect this has on a population in an environment of randomly placed food (which exist inside a certain boundary), is to prevent the individuals from ever leaving the area of food.

Interestingly, in tests where creatures are evolved from scratch in the benchmark environment, they rely on the patterns in the environment. Without randomness, they tend not to develop their sensory input data.

The more advanced problems are much more difficult to test, as they involve creature-to-creature interaction. The viable species evolved differed radically in their effectiveness, their main objective was to be able to decide how to prioritize their energy and reproductive needs. One strategy observed, was based on what colour the nearest object was, while another was to use it's internal energy status to decide what to do next.

Most of these viable species were happiest when grouped together quite closely, for obvious reasons, it is much easier to locate a mate if everything is close together. Attempts at breaking up the environment with barriers resulted in more diversification, due to the isolation of groups, but didn't alleviate this problem. Some of the best species evolved were the direct descendants of the "Blue foragers" described above, who adapted their foraging to mating.

One strategy, which emerged briefly at an early stage of the system, was an intriguing use of the abilities to signal (this may still be used, as signaling is very common). If a creature contained a high level of energy, it would activate its signal neuron. At the same time, surrounding creatures would attract to this signal colour and use it as an indication of mating potential. The result, that creature's save energy by not trying to mate with others with low energy.

The kind of animation produced by these species is very similar to the movements created by colonies of bacteria; early primitive creatures often travel in abrupt jumpy movements while more refined species smooth out their paths. Some results reflected those by Yeager [9], such as the grouping together of lazy creatures. It was possible to interactively develop the creatures indirectly by changing the environment while the simulation was being run. For example, if a population develops in close proximity, the distribution of food can be increased, making the environment more difficult and encouraging more active behaviour. As the simulation runs in real-time this adds a level of interaction, in being able to guide evolution.

8   Conclusions

The fascinating property of evolving systems is their ability to adapt and change to outside pressures. Any self-replicating process, which is open to mutation, will evolve. If the rules guiding the survival of these processes are manipulated, the results will change. Throughout the building of this system, the creatures have shown that they are able to adapt to, and make use of, errors and loopholes in the program code. It is this ability to survive and manipulate which is interesting.

The use of colour as a central property of recognition resulted in an emergent coherency of colour. Early experiments with 2D showed that that species could be distinguished easily and also made for good visual output. These ideas could be expanded further into full computer vision, where the texturing could also be used as a factor in object identification.

While the main purpose of this system was purely to produce animation, it is the processes, which define the animation, that hopefully make it worth watching. In general, the highly evolved behaviours provide much more interesting animation.

Theoretically is would be possible to breed creatures which exhibit a desired behaviour by tailoring the environment to suit. This way, an environment could be designed to create flocking systems or object avoidance. As shown by these early results, the solutions discovered would not always be obvious, and hopefully they would contain novel or interesting behaviours.

References

1. Alcock, J.: Animal Behavior: an evolutionary approach, Fifth Edition. Sinauer. Sunderland MA. (1993)
2. Dawkins, R.: The Blind Watchmaker. W.W. Norton, New York (1986)
3. Gould, S. J.: Wonderful Life. Hutchinson Radius, London (1990)
4. Langton, G.: [ed.] Artificial Life. The MIT Press, London. (1995)
5. Ray, T. S.: In press. A computational approach to evolutionary biology. "Advanced Mathematical Approach to Biology", Takeyuki Hida, [ed.]. World Scientific Publishing Co. Pte. Ltd., Singapore. Also, ATR Technical Report TR-H-176.
6. Sims, K.: Evolving Virtual Creatures, Computer Graphics (Siggraph '94) Annual Conference Proceedings, July, pp.15-22. New York: ACM Siggraph.
7. Sims, K.: Artificial Evolution for Computer Graphics, Computer Graphics (Siggraph '91 proceedings), Vol.25, No.4, July, pp.319-328.
8. Todd, S.,Latham, W.: Evolutionary Art and Computers, Academic Press. (1986)
9. Yaeger, Larry. Computational genetics, physiology, metabolism, neural systems, learning, vision, and behavior or PolyWorld: life in a new context. Artificial Life III, [Ed.] Christopher G. Langton, SFI Studies in the Sciences of Complexity, Proc. Vol. XVII, Addison-Wesley. Pp. 263–298.

Co-operative Improvement for a Combinatorial Optimization Algorithm

Olivier Roux, Cyril Fonlupt, and Denis Robilliard

Laboratoire d'Informatique du Littoral
Université du Littoral
BP 719
62228 Calais Cedex
FRANCE
Tel : +33-3-21-97-00-46
Fax :+33-3-21-19-06-61
e-mail :{roux,fonlupt,robillia}@lil.univ-littoral.fr

Abstract. These last years a new model of co-operative algorithm appeared, the model of ants colonies. This paper is dedicated to the integration of an ants colony's based co-operation method, in another algorithm, here research tabu, opposite the rough use of the computing power placed at the disposal on the current networks. The algorithms that we present are applied to the resolution of quadratic assignment problems (QAP).
Keywords : cooperative algorithm, ant colony, tabu search, QAP.

1 Introduction

The development of local area computer networks allows the use of this newly available computing power to solve combinatorial optimization problems. Often, this only consists in using the rough power offered by the network of machines (parallel execution of the same algorithm while launching as many copies of the algorithm as available machines). We propose an algorithm using a co-operative approach based on the model of the ants colonies presented for the first time by Dorigo *and al.* [10]. This approach was applied successfully to a number of algorithms (travelling salesman problem (TSP) [12], graph coloring problem [8], solving real function [3], vehicle routing problem [4], load balancing in telecommunications networks [22], . . .).

Our study is applied to the quadratic assignment problem [18] (QAP) which is known to be among one of most complex NP-hards problems [20]. In this article, we mainly compare two meta- heuristics : on the one hand, the PATS [29] (PArallel Tabu Search) known as a powerful method for the resolution of the QAP and, on the other hand, our new method based on an ants colony, so called ANTabu. We then validate our parallel co-operative choice versus the "brute force" parallel choice.

Note: The detailed characteristics of the parallel platform used for ANTabu (PVM) will not be detailed here.

C. Fonlupt et al. (Eds.): AE'99, LNCS 1829, pp. 231–241, 2000.
© Springer-Verlag Berlin Heidelberg 2000

2   The Quadratic Assignment Problem

The QAP is a combinatorial optimization problem, first described by Koopmans and Beckmann in [18]. It consists in finding the best assignment of n facilities to n locations at minimum cost. There are several QAP applications like arrangement of departments in hospitals, minimization of the total wire length in electronic circuits...

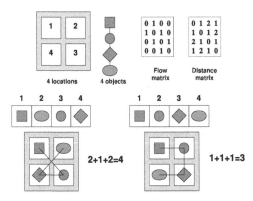

This trying sample (size=4) is represented by following elements: flow matrix, distance matrix, a random solution is presented on the left and on the right the best solution is described.

Fig. 1. resolution of a trying QAP example

This problem can be formally described as follows :

Given two symetric matrices D and C of size n, where D is the inter-city distance matrix and C the inter-city flow matrix, find a permutation π^* minimizing the objective function f :

$$\min_{p \in \Pi(n)} f(p) = \sum_{i=1}^{n} \sum_{j=1}^{n} d_{ij} c_{p_i p_j}$$

where :

p is a permutation.
p_i is the ith element of p.
$\Pi(n)$ is the set of all permutations of n elements.
$d_{i,j}$ is the distance between objects i and j (*i.e.* an element of D).
$c_{a,b}$ is the cost of flow between positions a and b (*i.e.* an element of C).

Thus a solution is an integer permutation, and an element of this permutation is an object number, its position within the permutation being the object location.

The QAP is a NP-hard problem and finding an ε approximation to this problem is known to be NP-complete [21], so problems of size larger than 20 are still considered as intractable for exact methods. Several heuristics have been proposed for finding near-optimum solutions to large instances, see for example [7], [23], [27], [2], [13], [1], [17], [24], [9], [19], [15], [29].

3   Ants Systems

The ants based algorithms were introduced in Marco Dorigo's PhD [10]. They are inspired by the observation that, using very simple communication mechanisms, an ant group is able to find the shortest path between any two points. During their trips a chemical trail (pheromone) is left on the ground. The role of this trail is to guide the other ants towards the target point.

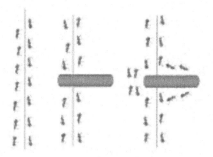

Fig. 2. Ants facing an obstacle

For one ant, the path is chosen accordingly to the quantity of pheromone. Furthermore, this chemical substance has a decreasing action over time, and the quantity left by one ant depends on the amount of food found. As illustrated in Fig. 2, when facing an obstacle, there is an equal probability for every ant to choose the left or right path. As the left trail is shorter and so requires less travel time, it will end up with a higher level of pheromone after many ants passages thus being more and more attractive. This fact will be increased by pheromone evaporation.

This subtle principle of communication has been used as a guideline for the design of heuristics dedicated to some combinatorial optimization problems, notably the TSP in [12,14,11], vehicule routing problems [4], load balancing in communication networks [6], real functions optimization [3], graph coloring problems [8],... Several works have also been targeted at the QAP, for example Stüzle's Min-Max algorithm [26], or Gamberdella et al.'s HAS-QAP [15].

234   O. Roux, C. Fonlupt, and D. Robilliard

3.1 From Ants to ANTabu

ANTabu is inspired by the natural model, but slightly deviates on some points. In nature, ants have a constructive behavior, that is they start from a point (the nest) and build a way throughout their tour. This behavior is found in a number of systems of ants colonies. However, from previous studies, this constructive approach seems less effective on the QAP, thus perturbation methods that modifies an existing solution are usually preferred, and it is our choice for ANTabu.

Another typical aspect found in ANTabu as well as in most ants colonies sytems, is the use of a "pheromon matrix", which purpose is the pooling of information coming from all the ants, and fed back to each of them. This global memory remembers solutions parts that are more frequently met in good solutions provided by the ants (hopefully these elements are probably components of a best solution). Evaporation is also simulated, thus we repeatedly decrease the values stored in the pheromon matrix, in order to avoid a too quick convergence, that could easily lead towards a bad local optimum.

In spite of the use of evaporation, ants can find themselves trapped around sub-optimal solutions. To avoid this, we also use a technique of diversification, which allows to reset ants with new solutions. This phenomenon can be seen as a metaphor of the behavior of natural ants, when they move their nest to face the rarefaction of food. In ANTabu, diversification is directed by a frequency matrix, thus new solutions are generated in parts of the search space that have not yet been visited.

The solutions provided by ANTabu are good, but still they can be improved by hybridization with local search heuristics. We have chosen a fast Tabu search for this purpose [16].

3.2 ANTabu Framework

ANTabu is based on the co-operation of two methods: a system of ants and an algorithm tabou [16]. The general scheme of our method is presented in the figure 3.2.

During the execution of ANTabu algorithm, the matrix of pheromones is updated using the solutions which are provided by the various ants. Thus the matrix is used cooperatively by the ants.

First of all, the matrix of pheromones is initialized with an arbitrary value proportional to a good solution obtained after some runs of local search starting from random solutions.

Each ant applies a certain number of swaps in the permutation, these swaps are guided via the matrix of pheromones. A swap consists in permuting the position of two pieces of equipment. The more there is pheromones for a possible assignment, the more it has chance to be selected. This operation provides a preliminary solution $\hat{p}$, the local search (Tabu) is then applied, and this gives a new solution $\widetilde{p}$. It should be noted that in our implementation, we limit the

Generate m (a number of ants) permutations p of size n
Initialization of the matrix of pheromones F
FOR $i = 1$ to I^{max} ($I^{max} = n/2$)
    FOR all the permutations p^k ($1 \leq k \leq m$)
        Building a solution :
            $\hat{p}^k = n/3$ transformations of p^k based on the matrix of pheromo-
        nes
        Local search:
            $\widetilde{p}^k = $ tabu search $(\hat{p}^k)$
        IF $\hat{p}^k < p^*$
            THEN the best solution found $p^* = \widetilde{p}^k$
        Update of the matrix of pheromones.
        IF $n/2$ iterations applied without improvement of p^*
            THEN Diversification

Fig. 3. ANTabu: general scheme

iteration number of the Tabu method in order to avoid a premature convergence towards a local optimum.

Solutions obtained after the local search stage, are used to update the matrix of pheromones. We proceed in two phases:

- a function simulating evaporation is applied (*i.e.* the values of pheromones in the matrix are decreased)

$$\forall i, j \in [1, n], \tau_{i,j}^g = (1 - \alpha)\tau_{i,j}^{g-1}$$

- each solution is used in the update of the matrix (*i.e.* one reinforces the assignments which are present in the solutions), this phase is detailed in the following section.

$$p \in P^g, \forall i, j \in [1, n], \tau_{i,j}^g = \tau_{i,j}^g + \frac{\alpha}{f(p)} \frac{f(p^-) - f(p)}{f(p^*)}$$

f is the function objective.
i is the position in the solution and j is its value.
$\tau_{i,j}^g$ is the value of the pheromone to g^{th} the iteration for the position i, j.
p^- is the worst solution found hitherto.
p^* is the best.
P^g is the whole set of solutions in the gth iteration.

Then ANTabu enters a phase of *intensification* if the best solution has just been improved, or a phase of *diversification* if no improvement has been obtained for several iterations.

3.3 Pheromones Management

The majority of the ants based algorithms take into account only the best solution found at each iteration for updating the matrix of pheromones (HAS-QAP[28], MAX-MIN[25], ...). In our method, each ant makes its own contribution as for AS[12]. This contribution is proportional to the quality of the solution found, and it is balanced by dividing it by the difference between worst and the best solution.

3.4 Diversification

Our strategy of diversification provides the algorithm with new solutions situated in parts of the search space that were not yet or only slightly explored. Opposite to traditional diversifications that simply use a random re-initialization of the matrix of pheromones, we use a memory, called *frequency matrix*, to keep a trace of all previous assignments, and this is used for diversification. During the generation of new solutions, each assignment is selected with a probability inversely proportional to the frequency value associated with its position (assignment selection is implemented via a roulette).

This strategy of diversification forces the system of ants to continue with solutions exhibiting a new structure.

4 Experimental Results

For all comparisons, we used ANTabu with a population of 10 ants. All the instances are extracted from the QAPlib[1] library, and the instances presented in the tables are those available in the articles we used as reference.

Note: one can differentiate mainly two classes of instances: regular instances (that includes mainly those generated by random, according to a uniform distribution) and irregular instances (that dominates in real life situations). The first class corresponds to a small flow of predominance while the second corresponds to a high flow value.

4.1 ANTabu for Short Executions

12 problems of QAPlib [5] were selected, some from the class of the irregular problems (bur26d, bur26b, chr25, els19, kra30a, tai20b, tai35b), others from the class of the regular problems (nug30, sko42, sko64, tai25a and wil50).

The parameters used for evaluation are: 10 iterations for each of the 10 ants, with a tabu limited to $5n$ iterations where N indicates the size of the problem considered, and a coefficient of evaporation $\alpha = 0.1$.

For this series of problems (table 1), the ANTabu algorithm obtains good performances in spite of the small iteration number.

---

[1] http://www.imm.dtu.dk/šk/qaplib/

Table 1. ANTabu on short problems and short runs. Values are given as the excess gap, *i.e.* (ANTabu result - Best known) / Best known, averaged on 10 iterations.

Problem name	Best known	ANTabu (excess gap)
bur26b	3817852	0.018
bur26d	3821225	0.0002
chr25a	3796	0.047
els19	17212548	0
kra30a	88900	0.208
tai20b	122455319	0
tai35b	283315445	0.1333
nug30a	6124	0.029
sko42	15812	0.076
sko64	48498	0.156
tai25a	1167256	0.843
wil50	48816	0.66

4.2   Validation of the Ants System Parameters

ANTabu differs from most ants systems by its diversification mechanism and its pheromon reinforcement procedure. So, what are the respective contributions of these two points?

Diversification. To validate the diversification method, we compared it with the basic strategy that consists in generating new random solutions. For this comparison, the two programs started a first diversification with the same solutions. This was done for thirty three problems of size ranging from 30 to 56, and results were averaged on thirty executions.

This comparison is very clearly in favour of the simple random diversification. This counter-performance seemingly tells us that elements in the least frequently found solutions, almost always form bad solutions that are very difficult to improve. This could lead one to suppose that good solutions are gathered in the search space, in the form of clusters. If it is the case, the solutions generated by our diversification mechanism are very distant from clusters of good solution, and thus harder to improve than random solutions.

Pheromone Update. To validate our implementation choice, we compared the use of contributions coming from all ants as opposed to only one ant (the best one), and a new formula of reinforcement of the pheromone. To run the comparisons, we implemented:

- a reference version with the Dorigo's update formula [12] for the best ant,
- a version with our update formula (Section 3.2) for the best ant,
- a version with the Dorigo's update formula [12] for all ants.

Benchmarks have been done on 25 instances of QAPlib with a size ranging from 30 to 40, results being averaged on thirty runs. The results are presented in table 2.

Table 2. Two new procedures of pheromone update compared with a version whithout both.

	Update contributions		Contributions from all ants	
	numbers of instances	proportion (percentage)	number of instances	proportion (percentage)
Number of results better than reference version	11	44.0	9	36.0
Number of results equal to reference version	11	44.0	11	44.0
Number of results lower than reference version	3	12.0	5	20.0

The results obtained validate these choices. In the large majority of cases, either the results were the same or were improved.

4.3   ANTabu vs PATS

We confronted our results with those obtained by the PATS algorithm [29]. This was done in order to validate our choice of co-operation and to check if our results were not simply due to the presence of a good local search. PATS is made of a set of distributed tabus exploiting the brute computing power of a network of workstations (including PC INTEL, Sun stations and an Alpha farm). The load of these workstations is monitored, and tabus are automatically launched on idle machines and withdrawn from those that are not idle any more (this is done automatically under the MARS platform [17]).

For this comparison, we studied instances belonging to two classes of problems:

- a problem with uniform matrices of distances and flow: tai100a
- some problems with a random matrix of flows: sko100a, sko100b, sko100c, sko100d, wil100

Among the instances suggested in [29], only those of size 100 for those PATS and ANTabu do not find the best solution known are presented in table 3.

The best solution found for PATS and ANTabu is the best solution for 10 executions. The duration time is given in minutes, and corresponds to the average for 10 executions on a network of 126 machines for PATS and on a network of 10 machines (SGI INDY) for ANTabu (thus PATS benefits from a higher computing power).

Table 3. Compared results from PATS and ANTabu. Best results are in boldface

	tai100a	sko100a	sko100b	sko100c	sko100d	wil100
QAPlib						
Best known	21125314	152002	153890	147862	149576	273038
PATS						
Best found	21193246	152036	153914	147862	149610	273074
Gap	0.322	0.022	0.016	0	0.023	0.013
Time	117	142	155	132	152	389
ANTabu						
Best found	21184062	152002	153890	147862	149578	273054
Gap	**0.278**	**0**	**0**	**0**	**0.001**	**0.006**
Time	139	137	139	137	201	139

The results show that ANTabu is better while using less computing power and time. It should be noted that the best solution is found for three of the four problems **sko100**.

In the PATS algorithm, tabus use random solutions, which they will try to improve throughout their execution. ANTabu also use random solutions at start, but share solutions parts in order to find interesting patterns, before applying tabu search. The parameters of tabu local search are similar in both PATS and ANTabu. Thus the difference in performance can only be explained by the use of co-operation, via the matrix of pheromones.

5   Conclusions

The co-operation between the virtual "ants" brings an undeniable profit, opposite to the rough use of parallel computing power. This pooling of information seems to be a good compromise between the division of all information and the speed of execution. And these results were obtained in spite of a diversification method that was proved to be ineffective in the end. The algorithm results are unequal depending on the instance, but always remain good. To bring more robustness, it could be interesting to have an algorithm that adapts itself to the nature of the problem which it tries to solve. This suppose that relevant measurements are available at hand: flow and distance dominance may seem interesting candidates.

References

1. V. Bachelet, P. Preux, and E. G. Talbi. Hybrid parallel heuristics : Application to the quadratic assignment problem. In *Parallel Optimization Colloquium*, Versailles, France, Mar 1996. POC96.

2. R. Battiti and G. Tecchiolli. The reactive tabu search. *ORSA J on Computing*, 6:126–140, 1994.
3. G. Bilchev and I. Parmee. The ant colony metaphor for searching continuous design spaces. *Lectures Notes in Computer Science 993*, pages 24–39, 1995. T. Fogarty (Ed.).
4. B. Bullnheimer, R.F. Hartl, and C. Strauss. A new rank based version of the ant system: A computational study. Working paper, University of Vienna, Austria, 1997.
5. R.E. Burkard, S.E. Karisch, and F. Rendl. Qaplib - a quadratic assignment problem library. *Journal of Global Optimization-10*, pages 391–403, 1997.
6. G. Di Caro and M. Dorigo. Antnet: A mobile agents approach to adaptive routing. Technical Report IRIDIA/97-12, IRIDIA, Universit Libre de Bruxelles, Belgium, 1997.
7. D.T. Connolly. An improved annealing scheme for the qap. *Eur J Op Res 46*, pages 93–100, 1990.
8. D. Costa and A. Hertz. Ants can colour graphs. *295-305*, Journal of the Operational Research Society(48), 1997.
9. V.-D. Cung, T. Mautor, P. Michelon, and A. Tavares. A scatter search based approach for the quadratic assignment problem. In *Proceedings of the IEEE Internatinnal Conference on Evolutionary Computation and Evolutionary Programming.*, pages 165–170, Indianapolis, USA, 1997. ICEC'97.
10. M. Dorigo. *Optimization, Learning and Natural Algorithms*. PhD thesis, Politecnico di Milano, Italy, 1992.
11. M. Dorigo and L.M. Gambardella. A study of some properties of ant-q. In *Proceedings of PPSN IV-Fourth -International Conference on Parallel Problem Solving From Nature*, pages 656–665, Berlin, Germany, September 22-27 1996.
12. M. Dorigo, V. Maniezzo, and A. Colorni. The ant system: Optimization by a colony of cooperating agents. *IEEE Transactions on Systems, Man, and Cybernetics*, 1-Part B(26):29–41, 1996.
13. C. Fleurent and J Ferland. Genetic hybrids for the quadratic assignment problem. *DIMACS Serie in Mathematics and Theoretical Computer Science*, 16, 1994.
14. L.M. Gambardella and M. Dorigo. Ant-q: A reinforcement learning approach to the traveling salesman problem. In A. Prieditis and S. Russell (Eds.), editors, *Proceedings of ML-95 - Twelfth International Conference on Machine Learning*, pages 252–260, Tahoe City, CA, 1995. Morgan Kaufmann.
15. L.M. Gambardella, E. Taillard, and M. Dorigo. Ant colonies for the qap. *Accepted for publication in the Journal of the Operational Research Society*, 1998.
16. F. Glove. Tabu search. *Journal of Computing*, Part I(1(3)):190–206, 1989.
17. Z. Hafidi, E-G. Talbi, and J-M. Geib. Mars : Un ordonnanceur adaptatif d'applications parallèles dans un environnement multi-utilisateurs. In *RenPar'8-8me Rencontres Francophones du Paralllisme*, pages 37–40, Bordeaux, France, Mai 1996.
18. T.C. Koopmans and M.J. Beckmann. Assignment problems and localisation of activities. *Economica*, 25(53-76), 1957.
19. P Merz and B. Freileben. A genetics local search to the quadratic assignment problem. In *Internationnal Conference on Genetic Algorithms, ICGA'97*, pages 465–472, New Lancing, Michigan, USA, 1997.
20. S. Sahni and T. Gonzales. P-complete approximation problems. *Journal of ACM-23*, pages 556–565, 1976.
21. S. Sahni and T. Gonzales. P-complete approximation problems. *Journal of ACM-23*, pages 556–565, 1976.

22. R. Schoonderwoerd, O. Holland, J. Bruten, and L. Rothbrantz. Ant-based load balancing in telecommunications networks. *Adaptive Behaviour*, 5,2:169–207, 1997.

23. J. Skorin-Kapov. Tabu search applied to the quadratic assignment problem. *ORSA Journal on Computing*, Vol.2(no. 1):33–45, 1990.

24. L. Sondergeld and S. Voß. *Meta-Heuristics : Theory and applications*, pages 489–502. Kluwer Academic Publishers : Boston/London/Dornecht, 1996.

25. T. Sttzle and H. Hoos. The max-min ant system and local search for the traveling salesman problem. In IEEE Press, editor, *Proceedings of ICEC'97-IEEE 4th International Conference on Evolutionary Computation*, pages 308–313, 1997.

26. T. Stützle. Max-min ant system for quadratic. Technical Report AIDA-97-04, AIDA, Darmstadt University of Technology, Computer Science Department, 1997.

27. E.D. Taillard. Robust taboo search for the quadratique assignment problem. *Parallel Computing 17*, 17:443–455, 1991.

28. E.D. Taillard and L. Gambardella. Adaptive memories for the quadratic assignement problems. Technical Report IDSIA-87-97, IDSIA, Lugano, Switzerland, 1997.

29. E-G. Talbi, Z. Hafidi, and J-M. Geib. Parallel adaptive tabu search for large optimization problems. In *MIC'97-2nd Metaheuristics International Conference*, Sophia Antipolis, France, Juillet 1997.

Landscapes and the Maximal Constraint Satisfaction Problem

Meriema Belaidouni[1] and Jin-Kao Hao[2]

[1] LGI2P, EMA-EERIE, Parc Scientifique Georges Besse, F-30000 Nîmes
[2] LERIA, Université d'Angers, 2 bd Lavoisier F-49045 Angers Cedex 01
{meriema@eerie.fr, Jin-Kao.Hao@univ-angers.fr}

Abstract. Landscape is an important notion in studying the difficulty of a combinatorial problem and the behavior of heuristics. In this paper, two new measures for analyzing landscapes are introduced, each of them based on the Hamming distance of iso-cost levels. Sampling techniques based on neighborhood search are defined in order to effect an approximation of these measures. These measures and techniques are used to analyze and characterize the properties of Maximal Constraint Satisfaction Problem random landscapes.

1 Introduction

Large combinatorial problems are often hard to solve since such problems may have huge configuration spaces. To tackle a large combinatorial problem, heuristics such as neighborhood search methods constitute one of the most powerful approaches. Though heuristics have proven to be very successful, there are few studies able to explain such a performance. The notion of landscape is among the rare concepts which assist in understanding the behavior of heuristics and in characterizing the difficulty of a combinatorial problem.

The concept of landscape was first introduced by Wright in 1932 [17]. Since then, the term has been re-used by several researchers, sometimes with different meanings [5], [11]. Numerous measures have been proposed for analyzing landscapes and for understanding their difficulty for heuristics. In what follows we introduce the notion of landscape and its measurement. Before entering into the detail, we give a definition of the concept of configuration space.

1. **Configuration Space**: Given a combinatorial problem P, a configuration space associated with a mathematical formulation of P is defined by a couple (S, f) where S is a finite set of configurations and f a cost function which associates a real number to each configuration of S. Only a small number of measures are available for this structure (configuration space). The minimum and the maximum are the two most common ones.
2. **Search Landscape**: Given a configuration space (S, f), a *search landscape* is defined by a triplet (S, v, f) where v is a neighborhood function which verifies $v : S \to 2^S - \{\emptyset\}$. This landscape, called an *energy landscape* in [5], can be considered as a "neutral" one since no search process is involved. This

C. Fonlupt et al. (Eds.): AE'99, LNCS 1829, pp. 242–253, 2000.
© Springer-Verlag Berlin Heidelberg 2000

landscape can be conveniently viewed as a weighted[1] graph $G = (S, v, f)$. Search landscapes have been the subject of a number of studies. Several measurements have been defined: first the number and distribution of *local minima*[2] [10], [13]; second *autocorrelation* - which quantifies the ruggedness of a landscape: *i.e.* the variation of the cost values between the neighbors in the graph [12], [16].

3. **Process Landscape**: Given a search landscape (S, v, f), a *process landscape* is defined by a quadruplet (S, v, f, ϕ) where ϕ is a search process. The process landscape represents a particular view of the neutral landscape (S, v, f) seen by a search process. This notion of landscape was first introduced in [11] for the purpose of studying the genetic operators of a genetic algorithm. *Autocorrelation of a random walk* [14] and *fitness distance correlation* (FDC) are well known measures of process landscapes [11].

Though these cost measures produce interesting and useful information about a landscape, it seems that they give only a partial picture. In particular, they fail to answer such important questions as: 1) How many configurations in the configuration space are there for a given cost? 2) Are the configurations in the configuration space scattered or gathered in specific areas? 3) What is the "accessibility" of one configuration from another?

Thanks to a measure called "density of states", recently proposed in [1], the first question is now answered. This measure, belonging to configuration space level, gives the number of configurations per cost value. In a previous study [4], we applied this state density measure to study the configuration space of random instances of the Maximal Constraint Satisfaction Problem (MAX-CSP). The estimation of state density reveals that the configurations of a given random MAX-CSP instance follow a gaussian distribution. This distribution constitutes an explanation of landscape difficulty for heuristics. Indeed, configurations are concentrated in certain cost zones attracting the heuristic and hindering its evolution towards less dense cost zones. Another important point is that this measure allows us to introduce a stratified model for configuration spaces (see Fig. 2 of Section 6). In this model, each cost value c is associated to the set of configurations having the cost c, which we call an iso-level. Simple distances based on cost values are also defined for the model.

In this paper, we try to answer the second and the third questions. Based on the above iso-cost level model, two new measures are proposed for landscapes: Hamming Distance Inside a Level (HDIL) and Hamming Distance Between Levels (HDBL). HDIL measures the similarity (or diversity) of configurations within an iso-cost level. This distance translates in a sense to the idea of the "width" of the landscape. Similarly, the distance between iso-cost levels (HDBL) measures the accessibility of configurations and reflects the "length" of the landscape. These two measurements together give another picture of a landscape.

---

[1] Note that the weights are defined on the nodes, not on the edges.

[2] $s \in S$ is a local minimum with respect to the neighborhood v if $\forall s' \in v(s)$, $f(s) \leq f(s')$.

At this stage, we notice that, there are some similarities between our work and the study concerning neutral networks recently reported in [2], though the two studies have quite different objectives and estimation techniques. Indeed, the model of neutral networks is close to the iso-cost level model. Measures such as neutral dimension and percolation index share some ideas with our HDIL and HDBL ones.

As in [4], we use the very general Maximal Constraint Satisfaction Problem as a test problem for our experiments. HDIL and HDBL measures are applied to landscapes corresponding to random MAX-CSP instances. Experimental results show that the various configurations of a given iso-cost level are a large distance apart. The results show also that the distance between iso-cost levels changes depending on whether high or low cost levels are considered.

This paper is organized as follows: Section 2 defines our new measures. Section 3 introduces sampling techniques. After a quick review of the MAX-CSP in Section 4, experimental results on random instances are given and discussed in Section 5. Finally, conclusions and future work are presented in Section 6.

2   Measures and Approximations

In this section, we will introduce two new measures for studying landscapes. We define *Hamming distance inside a level* (HDIL) which measures the average variation of distances for configurations having the same cost. Then we define *Hamming distance between levels* (HDBL) which measures the average distances to go from one configuration and cost to another. These two measures translate respectively the width and the length of a landscape.

2.1   Generalities

Given a configuration space (S, f), the neighborhood relation $N_1: S \to 2^S - \{\emptyset\}$ and the Hamming distance are defined as follows:

- **Neighborhood N_1**
  N_1 is defined as follows: two configurations $s(s_1 ..., s_n)$ and $s'(s'_1 ..., s'_n)$ involving n components (variables) are neighboring if they differ by a single value of a variable; more precisely, if and only if $\exists!(i, j)/s_i \neq s'_j$.
- **Hamming Distance**
  The Hamming distance associated with this neighborhood is:

$$d_H(s, s') = d_H(s', s) = \sum_{i=1}^{n} \delta(s_i, s'_i) \tag{1}$$

where $\delta(x, y) = \begin{cases} 1 \; if & x \neq y \\ 0 \; otherwise \end{cases}$

The Hamming distance $d_H(s, s')$ between two configurations s and s' counts the number of components having different values between s and s'.
- **iso-cost level**
  $C \subset S$ is an iso-cost level of cost $c \Leftrightarrow \forall s \in (C), f(s) = c$.

2.2 Hamming Distance Inside an Iso-Cost Level - HDIL

The distance inside a set A can be defined by the average distance between the elements of A.

$$D(A) = \frac{1}{|A^2|} \sum_{(s,s') \in C^2} d(s, s') \tag{2}$$

In particular, the *Hamming Distance Inside an iso-cost Level* C is defined by the distance inside a set with $A = C$ and $d = d_H$. It corresponds to a measure of landscape width. This distance, denoted by $D(C)$, represents the diversity (or similarity - if one replaces δ by $\delta' = 1 - \delta$) of configurations of C. The HDIL measure differs from autocorrelation in measuring the variation of distance for a fixed cost while autocorrelation measures the cost variation for a fixed distance. This measure has two advantages: it concerns all costs areas and it is not dependent on instance size.

2.3 Hamming Distance between Iso-Cost Levels - HDBL

The Hamming distance between two sets A and A' can be defined by the average distance between pairs of elements of A and A'.

$$D(A, A') = \frac{1}{|A \times A'|} \sum_{(s,s') \in A \times A'} d(s, s') \tag{3}$$

We define the *Hamming Distance Between iso-cost Levels* C and C' by the Hamming distance between sets with $A = C$, $A' = C'$ and $d = d_H$. Below, this distance is denoted by $D(C, C')$. This measure can sometimes yield very large distances and has little meaning if it is used to study the behavior of search algorithms. The "vertical" Hamming distance between two levels, defined in the following formula, seems to be more relevant.

$$D_{CC'} = \frac{1}{|C|} \sum_{s \in C} d(s, s') \tag{4}$$

where $s' \in C'$ is a configuration reachable from s by a search process ϕ. Thus, this distance concerns configurations 'linked' by a search process. The measure $D_{CC'}$ informs us of the distance between two successive levels C and $C + \delta C$. Using it, it is possible to know the potential ancestor of a configuration. Note though that the measure $D_{CC'}$ is different from FDC since $D_{CC'}$ concerns distance variations while FDC concerns the relationship between distance variation and cost variation. Moreover, in $D_{CC'}$, the distance is measured with respect to a set of configurations, whereas FDC has a single reference configuration.

2.4 Estimators

Because of the large number of configurations of a given cost value c, it is difficult to calculate the distances $D(C)$, $D(C, C')$ and $D_{CC'}$ exactly. We can though approximate them by using sufficiently large and representative subsets $E \subset C$ and $E' \subset C'$. The corresponding estimators $\hat{D}(C)$, $\hat{D}(C, C')$ and $\hat{D}_{CC'}$ are defined as follows:

$$\hat{D}(C) = \frac{1}{|E^2|} \sum_{(s,s') \in E^2} d_H(s, s') \tag{5}$$

$$\hat{D}(C, C') = \frac{1}{|E \times E'|} \sum_{(s,s') \in E \times E'} d_H(s, s') \tag{6}$$

$$\hat{D}_{CC'} = \frac{1}{|E|} \sum_{s \in E, \psi(s) \in E'} d(s, \psi(s)) \tag{7}$$

where $\psi(s)$ is the first configuration of E' encountered by the search process ψ.

3   Sample Construction

Sample construction is a decisive stage for determining approximation validity. As our measures are related to the level C of a cost c, we need a sampling process which is able to reach configurations of the cost c. Moreover, the sampling process must be able to reach sufficiently diversified configurations in order to constitute a representative sample E of C. Theoretically, random search might be a suitable sampling process. However, random search is not viable in practice since it is unable to reach configurations in low cost areas (for minimization). In this work, we adopt neighborhood search methods including Metropolis, Simulated Annealing and Tabu Search to reach low cost areas. These methods have been successfully used to resolve the MAX-CSP problem [7], [9]. Two sampling techniques are proposed; the first may use any "neighborhood heuristic", while the second is based on the Metropolis algorithm.

3.1   Sampling with Re-run

Sampling with re-run consists in building a representative sample by keeping only one configuration per run (or execution) of any neighborhood heuristic. Thus, to build a sample E, at least $|E|$ runs are needed. A single run of any neighborhood search process does not gather a sufficient number of configurations of a given cost. In order to increase this number we run the heuristic (Tabu, Simulated Annealing and Metropolis) the required time to obtain a sample of sufficient size.

This simple technique does not require any tuning (except to reach the wanted cost value). Moreover, as configurations of the sample come from different runs, it insures configuration independence. However, it may be very time consuming.

3.2   Sampling without Re-run

Sampling without re-run consists in building a sample of a given cost c in only one run (execution) of a given "neighborhood heuristic". This task is not easy because, most of the time, neighborhood search methods generate only a few configurations for a given cost value in any one run.

However, there exists a neighborhood heuristic for which a particular phenomenon occurs. As shown in [4], Metropolis applied to MAX-CSP random instances stagnates around an average cost value which depends only on the temperature used. Fig. 1 shows the phenomenon with different temperatures $T = \{35, 15, 5, 0.5\}$. For example, for $T = 0.5$, 0.075% of the configurations met in one run have a cost of 32. Therefore, in order to yield a sample of size K with a cost of 32, it is sufficient to run the Metropolis process with $T = 0.5$ for N iterations ($K = 0.075\% \times N$).

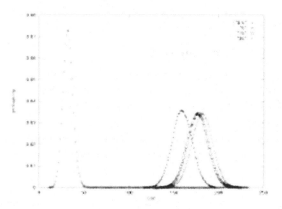

Fig. 1. Metropolis for different temperatures

The technique of sampling without re-run can be very useful: it is a fast method (one run is sufficient). However, the sampled configurations are not completely independent of each other.

In what follows, we define the Maximal Constraint Satisfaction Problem - a general optimisation formalism. We then go on to apply the measures proposed here to comparing binary MAX-CSP landscapes.

4   Maximal Constraint Satisfaction Problem (MAX-CSP)

4.1   Problem Definition

The MAX-CSP can be conveniently defined by means of the notion of *constraint network*. A constraint network CN is a triplet $< V, D, C >$ where:

- V = $\{V_1, V_2...V_n\}$ is a finite set of variables;

- D = $\{D_1, D_2...D_n\}$ is a finite collection of value domains associated with the variables;
- C = $\{C_1, C_2...C_n\}$ is a set of constraints, each being a subset of the Cartesian product of the domains of some variables specifying allowed (or forbidden) value combinations.

Given such a constraint network $CN < V, D, C >$, the MAX-CSP becomes the finding of an assignment of the values of D to the variables of V such that the number of satisfied constraints is maximized [6]. In practice, the equivalent minimization version is often used *i.e.*, instead of maximizing satisfied constraints, one minimizes the number of violated constraints. The MAX-CSP can be formulated as a couple (S, f) where S is the set of all possible assignments of values of D to the variables of V and f is the number of unsatisfied constraints of C. Note that the classical Constraint Satisfaction Problem (CSP) is a special case of MAX-CSP for which the optimal cost $f^* = 0$.

5   Experimentation

5.1   Random Instance Generation

Test instances used in this work correspond to random binary constraint networks generated according to a standard model [15]. A network class is defined by $< n, d, p_1, p_2 >$ which has n variables, d values per variable, $p_1.n.(n-1)/2$ constraints taken randomly from $n.(n-1)/2$ possible ones (p_1 is called the density), and $p_2.d^2$ forbidden pairs of values taken randomly from d^2 possible ones for each constraint (p_2 is called the tightness). For each given class $< n, d, p_1, p_2 >$, different instances can be generated using different random seeds.

A constraint network may be either under-constrained or over-constrained. A phase transition in solubility occurs in between when the network is critically constrained [15]. Under-constrained networks tend to be easily satisfiable (cost $f^* = 0$). Over-constrained networks are usually unsatisfiable (cost $f^* > 0$). Critically constrained networks may or may not be satisfiable and in addition are usually hard to solve from a satisfaction point of view. These different regions are characterized by a factor called *constrainedness* [8]:

$$\kappa = \frac{n-1}{2}p_1 \log_d(\frac{1}{1-p_2})$$

$\kappa = 1$ delimits under- ($\kappa < 1$) and over- ($\kappa > 1$) constrained networks. Networks with $\kappa \approx 1$ corresponds to critically-constrained ones. For the purpose of this work, we use over- ($\kappa > 1$) or critically ($\kappa \approx 1$) constrained networks.

Two distance measures are applied to the search landscape (S, v, f) of random MAX-CSP instances. HDIL and HDBL are calculated and analyzed for various cost values. The random instance $< 100.10.15.25 >$[3] is used as an example. This instance has a known optimal value of $(f^*)=0$ [9] and a center cost level $C_m = 186$ [4]. The MAX-CSP landscapes are supposed to be isotropic; the calculations confirm this assumption.

---

[3] This instance is generated with a random seed equal to 3.

5.2 Hamming Distance Inside an Iso-Cost Level

The purpose of the following experiment is to estimate Hamming distances in various levels of the instance under study $< 100.10.15.25 >$. Sampling with re-run, Simulated Annealing (SA) and Tabu Search (TS) is used to sample given cost configurations. Preliminary tests were carried out to determine the number of configurations in a sample E. It has been found that 5000 configurations are sufficient to make the statistics stable. Therefore, both SA and TS are run 5000 times for each of the fixed costs of the target cost level. Table 1 gives the TS results for different cost levels running from the average instance cost 186 to the cost 12. For each level, we count the number of executions (called size) among the 5000 that meets the level in question and record its Hamming distance. The experiments are repeated 10 times. The first thing one can notice is that some re-runs failed to reach wanted cost configurations. For example, for 5000 executions, TS reaches the cost value 183 only 332 ± 19 times. However, size increases when cost decreases; it reaches 4932 ± 4 for cost level 12. As far as Hamming distances are concerned, large magnitudes (Table 1) are obtained. Similar results are observed for costs lower than 12.

In the same way, Table 2 gives the results obtained with SA. Once again, we observe large magnitudes for Hamming distances. Similar results are obtained by sampling without re-run using Metropolis. The Hamming distances are very close to those generated by Simulated Annealing Table 2. Similar results are observed for costs lower than 12.

At this stage, some conclusions can be made. The measure of HDIL discloses some interesting aspects of the landscape for random instances. A large HDIL width within a level shows that configurations are not clustered in a single group (which does not mean that they do not belong to large clusters). One implication of this result is that a single configuration cannot represent a whole level. Another point is that HDILs obtained over one run are always smaller than those obtained by our sampling method; therefore our results can be considered as upper bounds on the instance in question. Moreover, Hamming distances obtained by Simulated Annealing and Metropolis are smaller than those obtained by Tabu as showed in Table 1 and Table 2. Simulated Annealing and Metropolis reach the desired (low) cost levels more frequently. Tabu therefore explores a larger low cost area of the landscape. In addition to these considerations, one should notice that though Tabu search is faster than Metropolis and Simulated Annealing, all these calculations are time consuming (Table 1 and Table 2).

Table 1. Hamming distance inside an iso-cost level (HDIL) by sampling with re-run (Tabu)

cost	Tabu(30)					
	12	32	82	125	159	183
size	4932 ± 4	4766 ± 13	1902 ± 31	1219 ± 25	906 ± 19	332 ± 19
distance	0.862 ± 10^{-3}	0.882 ± 10^{-3}	0.895 ± 10^{-3}	0.898 ± 10^{-3}	0.899 ± 10^{-3}	0.899 ± 10^{-3}

Table 2. Hamming distance inside an iso-cost level (HDIL) by sampling with re-run (Simulated Annealing)

	Simulated Annealing					
cost	12	32	82	125	159	183
size	4738	4260 ± 30	3465 ± 26	2948 ± 30	2519 ± 36	1303 ± 42
distance	0.8188	0.851 ± 10^{-3}	0.886 ± 10^{-3}	$0.896 \pm 10.^{-3}$	0.899 ± 10^{-3}	0.899 ± 10^{-3}

5.3 HDIL on other Instances

We have measured Hamming distance inside an iso-cost level of different random MAX-CSP instances (Table 3). These instances have sizes ranging from 100 to 300 variables with κ around 1. For each instance, we give the near optimal cost value (f^*), the central cost value (C_m), the Hamming distance inside a level of cost C_m $(D(C_m))$. We apply sampling with re-run using Tabu search - to approximate HDIL.

We observe from Table 3 that for all instances, $HDIL$ is about 0.9, which corresponds to consistently large diversity. This consistency may have something to do with random generation. Even if this value may be conditioned by the approximation technique used, it seems possible that the original value is large.

Table 3. HDIL on other instances

	κ	f^*	C_m	$D(C_m)$	$D(\frac{C_m-f^*}{2})$	$D(f^*)$	D_{f^*,C_m}
(a) 100.10.15.25	0.93	0	186	0.89	0.89	0.86	52
(b) 100.10.20.25	1.24	19	248	0.90	0.89	0.85	54
(c) 100.15.10.45	1.64	11	223	0.93	0.93	0.90	55
(d) 100.15.20.30	1.3	25	297	0.93	0.93	0.90	60
(e) 100.15.30.20	1.22	18	297	0.93	0.93	0.89	56
(f) 200.20.20.15	1.08	21	597	0.94	0.94	0.93	119
(g) 200.20.18.14	0.9	0	501	0.95	0.94	0.93	112
(h) 300.20.10.18	0.99	14	807	0.95	0.94	0.93	168
(i) 300.30.16.12	0.9	4	861	0.96	0.96	0.96	172

5.4 Hamming Distance between Iso-Cost Levels

Hamming distance between levels expresses the length of a landscape. We would like to answer questions like: can one estimate the distance between two levels? If yes, can one draw up a distance map? Or are levels at equal distances apart? Can one know the potential ancestors of a given configuration? Do these distances change with the search heuristic?

The goal of this experiment is to compute the Hamming distances between levels for the descent using sampling with re-run. Let us recall that the descent process begins with a random initial configuration s_0, then at each iteration, moves to a better neighbor. The process stops when a local optimum is reached. We run 1000 executions of this strict descent algorithm and we compute distances between levels in the form of a map of distances using formula (7). We analyze the results on two levels. Area (A) in the left part of Table 4 corresponds to a high cost area (average cost around $C_m = 186$). Area (B) (at right of the same table) corresponds to a low cost area. The results are interpreted as follows. For $\hat{D}_{179,177}$, if the search process wants to go from a configuration of cost 179 to one of cost 177, a distance of 1.3 is necessary. The results shows that for a given cost value variation, configurations are further apart in lower cost levels. For instance, by comparing the distances $\hat{D}_{179,177}$ and $\hat{D}_{39,37}$, which both have a cost variation of 2, we observe that $\hat{D}_{179,177} < \hat{D}_{39,37}$. This comparison is valid for other distances between levels.

The increase of Hamming distance between levels shows the flat structure of this landscape in low cost areas.

Table 4. Distance map between levels for high and low cost areas

$\hat{D}_{C,C'}$	177	178	179	180	181	182	183
179	1.3	1	0	1	1.4	1.7	2.2
180	1.8	1.3	1	0	1	1.3	1.7
181	2.2	1.8	1.4	1	0	1	1.4

$\hat{D}_{C,C'}$	37	38	39	40	41	42	43
39	1.9	1	0	1	1.8	2.8	3.6
40	2.8	1.9	1	0	1	1.9	2.7
41	3.6	2.8	1.8	1	0	1	1.9

5.5   HDBL on other Instances

Similar calculi have been carried out for instances (a) to (i) of Table 3. Given the fact that the computing of an HDBL distance map is quite long, we limited our computing to the distance between a cost level C_m and the cost level f^*. We used the sampling technique with re-run with the following "reversed" random walk: one begins with an optimal solution and then makes a random walk at each step.

Again, the HDBL results obtained look rather similar to those for instances with the same number of variables. This consistency may have something to do with random generation. At the same time, $C_m - f^*$ is not the same for all instances. Therefore, $\frac{D_{f^*,C_m}}{C_m - f^*}$ may be used to compare instances. However, our actual sampling processes are not fast enough to yield a precise approximation of this measure.

6   Conclusions and Future Work

In this paper, we have proposed two measures - called *Hamming Distance Inside a Level or HDIL* and *Hamming Distance Between Levels or HDBL* - to analyze search landscapes (S, v, f). Both distances concern links between components of configurations. However, Hamming distance inside a cost level quantifies the diversity of different configurations or the width of a landscape whereas Hamming distance between cost levels translates the accessibility of configurations or the length of the landscape. Hamming distances inside and between levels can be represented as follows Fig. 2: .

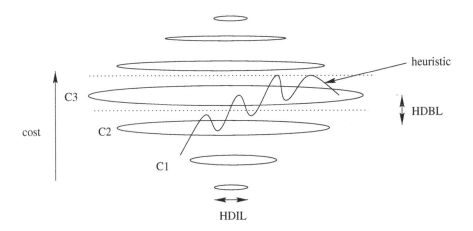

Fig. 2. Iso-cost level model for configuration space (S, f)

To estimate these distances, sampling techniques based on neighborhood searching were developed. Notice that, depending on the required precision, the approximation of HDIL and HDBL may be time consuming.

These measures have been applied to a number of random MAX-CSP instances. Experimental results show that, within a level, the configurations have a large distance (about 0.9 for a maximum of 1). This distance remains stable for different cost levels.

Experiments with HDBL disclose that the distance between two different high-cost levels is smaller than the distance between two low-cost levels. Moreover, these distances change for different instances.

Such information may be used as a criterion to compare the structure of different instances. The information may also help to predict the possible adaptation of heuristics on the landscape.

Currently, we are working on two points. First, we are studying other search landscapes using the measures proposed in this paper. Second, we are working on the *process landscape* (S, v, f, ϕ). In particular, we are studying the characterization of the Metropolis process [3]. This study is based on statistical features obtained with Metropolis sampling at different temperatures. It allows one to associate a curve with each process landscape, and consequently to compare process landscapes.

Acknowledgments. The authors highly appreciate the valuable and detailed comments of the anonymous reviewers.

References

1. Asselmeyer, E., Rosé H., Korst, J.: Smoothing representation of fitness landscape - The genotype phenotype map of evolution. Biol. Systems, (1995)
2. Barnett, L.: Evolutionary dynamics on fitness landscapes with neutrality. MSc Dissertation. http://www.lionelb.cogs.susx.ac.uk, (1997)
3. Belaidouni, M., Hao, J.K.: A measure of combinatorial landscape difficulty for the Metropolis algorithm. Submitted, (1999)
4. Belaidouni, M., Hao, J.K.: Search space analysis of the maximal constraint satisfaction problem (in French). Proc. Of 5th French Workshop on Practical Solving of NP Complete Problems. 58-71, (1999)
5. Catoni, O.: Rough large estimates for simulated annealing: application to exponential schedules. The Annals of Probability. 20(3): 196-208, (1997)
6. Freuder, E.C., Wallace, R.J.: Partial Constraint Satisfaction. Artificial Intelligence. 58(1-3): 21-70, (1992)
7. Galinier, P., Hao, J.K.: Tabu search for maximal constraint satisfaction problems. LNCS 1330: 196-208, (1997)
8. Gent, I.P., MacIntyre, E., Prosser, P., Walsh, T.: Scaling effects in the CSP phase transition. Proc. of CP95, 70-87, (1995)
9. Hao, J.K., Pannier, J.: Simulated annealing and Tabu search for constraint solving. Artificial Intelligence and Mathematics IV, (1998)
10. Hertz, A., Jaumard, B., De Aragao, M.P.: Local optima topology for the k-coloring problem. Discrete Applied Mathematics 49: 257-280, (1994)
11. Jones, T., Forrest, S.: Fitness distance correlation as a measure of problem difficulty for genetic algorithms. Proc. of the 6th International Conference on Genetic Algorithms. 184-192, (1995)
12. Kauffman, S. A.: Adaptation on rugged fitness landscapes. Lectures in the Sciences of Complexity. SFI Studies in the Sciences of Complexity, ed. D. Stein, Addison-Wesley Longman, 527-618, (1989)
13. Kirkpatrick, S., Toulouse, G.: Configuration space analysis of travelling salesman problem. J. Physics. 46: 1277-1292, (1985)
14. Manderick, B., de Weger, B., Spiessens, P.: The genetic algorithm and the structure of the fitness landscape, PPSN III, (1994)
15. Smith, B.M.: Phase transition and the mushy region in constraint satisfaction problems. Proc of ECAI 94. 100-104, (1994)
16. Weinberger E. : Correlated and uncorrelated fitness landscapes and how to tell the difference, Biol. Cybern. vol. 63: 325-336, (1990)
17. Wright, S.: The roles of mutation, inbreeding, and selection in evolution. Proc. of the Sixth Congress on Genetics. Vol.1: 365, (1932)

Synthetic Neutrality for Artificial Evolution

Philippe Collard, Manuel Clergue, and Michael Defoin-Platel

Laboratoire I3S, CNRS - UNSA
2000 route des lucioles, 06410 Biot, FRANCE
e-mail:{pc,clerguem,defoin}@i3s.unice.fr

Abstract. Recent works in both evolution theory and molecular biology have brought up the part played by selective neutrality in evolution dynamics. Contrary to the classical metaphor of fitness landscapes, the dynamics are not viewed as a climb towards optimal solutions but rather as explorations of networks of equivalent selective genotypes followed by jumps towards fitter networks. Although the benefit of neutrality is well known, it is hardly exploited in the genetic algorithm (GA) field. The only works about this subject deal with the influence of the inherent neutrality of a fitness landscape for evolution dynamics. In this paper, we propose a very different approach which consists to introduce "handmade" neutrality into the fitness landscape. Without any hypothesis about the inherent neutrality, we show that a GA is able to exploit new paths through the fitness landscape owing to the synthetic neutrality.

1 Introduction

A theory of the biological evolution at molecular level based on neutrality was first proposed by Kimura ([9]). According to this work, the most part of changes during evolution arises from a random drift at genotype level rather than from the effect of selective pressure. The neutral theory supposes firstly that there are sets of selectively equivalent genotypes and secondly that it is possible to move through those sets using selectively neutral mutations. Kimura is interested only in the influence of the neutrality for the dynamics in flat adaptive landscapes. Further, many simulations based on the RNA secondary structure ([6,8]) shown that neutrality plays an important role in the evolution dynamics at a molecular level.

Thus, the neutral theory explains evolution by diffusion processes through neutral networks (i.e. sets of connected genotypes that are selectively equivalent). This diffusion increases the opportunity to find a part of the genotype space in which the selectively dominant network is closed to a more adapted network. More generally, there are a lot of interlaced networks; moving through a network, a population generates some mutants out of the network and then explores new phenotypes. A transition between two phenotypes appears in regions where networks become closer. This makes the classical vision of a rugged adaptive landscape, with peaks and valleys, no more relevant ([1]).

C. Fonlupt et al. (Eds.): AE'99, LNCS 1829, pp. 254–265, 2000.
© Springer-Verlag Berlin Heidelberg 2000

More recently, some adaptive landscapes owning a high neutrality level have been synthesized in the fields of cellular automata ([7]), sequential dynamic systems ([2]), and genetic algorithms ([1]).

Many fitness landscapes from genetic algorithms literature own an high level of neutrality, for example, the Royal Road functions or unitation based functions (e.g. the "trap" function). Those landscapes have been generally designed to study one kind of difficulty : deceptivity, building blocks processing, etc. So, it is possible to build highly neutral fitness landscapes that are either GA-difficult or GA-easy. There is *a priori* no correlation between GA-difficulty and neutrality. But, can we say that a GA is not able to exploit the neutrality of a fitness landscape ? In this paper, we give a partial answer to this question. Rather than evaluating the neutrality level of a given landscape or designing fitness landscapes with controllable neutrality level, as for example, the NKp landscapes family ([1]), we propose to introduce, for any landscape, a structured neutrality. In this paper we show how the exploitation of such a neutrality can help a GA to find the way towards the global optimum. We propose a first step that consists in building some *ad hoc* fitness landscapes in which we think that the GA can exploit the synthetic neutrality. Then, we will study the populations dynamics in those landscapes. Our aim is twofold: firstly, we propose a new technique based on the synthetic neutrality concept; then, we build a family of adaptive landscapes owning a given neutrality level and we design some visualization and measure oriented tools to study the populations dynamics.

Finally, we want to say some words about the methodology. We think that we have to begin our studies with *ad hoc* landscapes because it is easier to define such landscapes than more realistic ones. We also notice that there are many things to understand about neutral dynamics in simple models and that we are not ready to study more complicated models in which the parts played by neutrality and selection could not be clearly discerned.

2   Synthetic Neutrality

We've already reminded the role of neutrality in biological evolution. This led some authors to study the dynamics of GA from a neutral point of view (e.g. [1]). They analyzed the influence of the inherent neutrality of particular adaptive landscapes on the GA dynamics. Here, we propose a way to artificially introduce neutral networks into fitness landscapes.

2.1   Minimal Synthetic Neutrality

The basic idea of the synthetic neutrality is to create an extended genotype space, on which the GA works, and then, an application from this new space to the basic genotype space on which the evaluation function works.

Let $\Omega = \{0, 1\}^\lambda$ be the basic genotype space. In a minimal way, the extended genotype space, $\langle \Omega \rangle$ is defined as $\{0, 1\} \times \Omega$. It corresponds to the adjunction of an additional bit to the chromosomes from Ω. This new bit, called the *meta-bit*, controls the interpretation of the rest of the binary string.

We define the application T from $\langle \Omega \rangle$ to Ω as :

$$\forall \omega \in \Omega \ , \ T(0\omega) = T(1\overline{\omega}) = \omega$$

where $\overline{\omega}$ is the binary complement of ω. Let F be the evaluation function defined over Ω. From the point of view of $\langle \Omega \rangle$, the algorithm works on a new fitness landscape defined by the function $\langle F \rangle = F \circ T$ over $\langle \Omega \rangle$. The following relation can be established :

$$\forall \langle \omega \rangle \in \langle \Omega \rangle, \langle F \rangle(\langle \omega \rangle) = \langle F \rangle(\overline{\langle \omega \rangle}) = F(T(\langle \omega \rangle))$$

Thus we get, without any hypothesis about the inherent neutrality of Ω, two genotypes from $\langle \Omega \rangle$ with the same adaptive value, so selectively equivalent.

Now, we have to show how such genotypes could be connected in a same neutral network. Indeed, classical mutation is not adequate here since those genotypes are at maximal Hamming's distance $(\lambda + 1)$ of each other because of their complementarity. This leads us to introduce a neutral operator of mutation, also called mirror operator (according to its effects on the genotype or on the phenotype). This operator transforms a chromosome from $\langle \Omega \rangle$ to its binary complement. According to the above relation, this transformation doesn't induce any modifications on the phenotype. With this operator, two complementary chromosomes (that are selectively equivalent) are linked by a neutral network, since they are only one neutral mutation apart.

Such synthetic neutrality is said minimal, according to the size and the dimension of the induced networks. Indeed, in $\langle \Omega \rangle$, there are 2^λ networks which the size is 2. Their dimension (number of possible neutral mutation from a point) is equal to 1.

Despite its minimality, this synthetic neutrality (MSN) has interesting properties. The first one is that neutral mutations and classical mutations (that infer a change in the phenotype) are independent from each other. The neutral mutation rate is set on an high value (around 0.01), while the classical mutation is still fixed at its usual value (around $1/\lambda$). Thus, we can get an important ratio between neutral and classical mutation rates, as in neutral networks of high dimensions. The second property of the MSN affects the neighborhood of a neutral network. This neighborhood contains some chromosomes that could be very distant in $\langle \Omega \rangle$, as in Ω. Indeed, the introduction of synthetic neutrality alters the topology of Ω. This property of the MSN reminds the drifts in larger neutral networks that allow huge jumps through the search space.

2.2   Order n Synthetic Neutrality (SN$_n$)

Following the same idea, it is possible to synthesize a higher level of neutrality, according to the size and the dimension of neutral networks. Indeed, we can introduce several meta-bits, each influencing a part of the chromosome. Each part with its own meta-bit is called a block. The space of chromosomes becomes the space $\langle\Omega\rangle_n = \{0,1\}^n \times \Omega$ with n meta-bits. For $n = 0$, this is a standard GA, while the case $n = 1$ stands for the minimal synthetic neutrality. The maximal bound for n is λ, since this is uninteresting to have more blocks than bits in the chromosomes. Connections between equivalent genotypes are created by the partial mirror operator. This operator is selectively neutral and allows to complement just one block. The number of neutral networks stays equal to 2^λ, but their size and dimension become respectively 2^n and n. The two properties of the minimal synthetic neutrality are still true. The synthetic neutrality is structured and affects the search space topology by bringing closer initially distant chromosomes. For example, let us consider chromosomes 0010010011 and 1101110001. The Hamming's distance between them is high in Ω ($d_H = 6$). On the other hand, in $\langle\Omega\rangle_2$, they may be closer. Indeed, in this space, the distance may be as small as 2.

3 Populations Dynamics and Synthetic Neutrality

Our goal, in this section, is to show how the synthetic neutrality can affect the dynamics of the populations.

3.1 Methodology

It's difficult to well understand how a population can exploit neutrality during the evolution. Indeed, classical and neutral evolution behaviors are often concomitant. That's why we propose to bring in light the exploitation of neutral networks in experimental situations specifically designed to promote it. For this purpose, we build particular GA-hard functions. Then we study the dynamics of the population.

3.2 Adaptive Landscapes

A fitness function gives the distribution of adaptive values along the search space, also called the fitness landscape ([10]). The dimensions of such search space are often high, thus we propose to design fitness landscapes in the plan fitness / distance to the optimum. Let's notice that this projection implies that all genotypes at a given distance from the optimum own the same fitness. The global optimum is set, without loss of generality, to the genotype 0^λ (0.....0).

During our experimentations, we have considered first and second order synthetic neutrality oriented landscapes. When we design such landscapes, according to the theoretical model, we hope for being able to predict spatial and temporal circumstances of the appearance of particular events. There are at least two reasons that prevent us from a priori determining qualities of a given landscape in a neutral context. Firstly, as Barnett says ([1]), there is an independence between ruggedness measure of a landscape (given by the autocorrelation function) and the neutrality level. Secondly, it is hard to predict how transformations induced by synthetic neutrality will be exploited during evolution.

3.3   Visualization Tools

Visualize the dynamics of the population in its globality is quite impossible because search space dimensions are generally too huge ([5]). So, we have to visualize these dynamics from a more restricted, then more ambiguous point of view. Nevertheless, varying points of view can remove many ambiguities.

Another problem is that these dynamics seem to be determined by many parameters such as : fitness landscapes characteristics, population polymorphism, "birth" of a dominant individual, *etc.* That is why we developed several tools, each one designed to observe particular features.

3.3.1   Macroscopic Visualization

Tools presented here allow to study the mean properties of a population.
Centroid trajectory :
Population centroid is the point representing the mean fitness and the mean distance to the global optimum plotted in the corresponding plan. By determining this point at each generation, we can get a trajectory which is a good reflect of the population dynamics through the associated landscape. We're able to study this trajectory in the basic search space, as well as in the extended search space.
Distance to the optimum per block :
When we have studied the order 2 neutrality, we were interested in blockwise evaluation of genotypes. This means that global fitness of a genotype is given by fitness function computed on each block. So the distances to the optimum of each block are relevant measures. Computing these two distances allows to draw both a three dimensional trajectory (mean distance from the block 1, mean distance from the block 2, mean fitness) or the time evolution of each block distance.

3.3.2   Microscopic vs. Macroscopic Visualization

Tools presented here allow to study the relations between one individual and its population. The main objective is to show how some particulars events could influence the global populations dynamics.
Genotypes evolution :
The population evolution might be influenced by the coming out of a particular genotype. In order to verify this hypothesis, we propose to plot the added up proportions (computed at each generation) of all the genotypes of the population. Computational constraints force us to limit the genotype size to 64 bits. In order to observe possible moves of population through order 1 neutrality networks, we choose to draw those proportions according to two significant schemata : $0\#$ (positive ordinate) and $1\#$ (negative ordinate), with $\# \in \{0,1\}^{\lambda}$.
Mean and maximum fitness evolution :
This is a classical tool that bring in light the relations between the fittest genotype and the mean fitness of the population.
Relative trajectory of the centroid :
This relative trajectory is nearly similar the one previously described. The main difference is that the centroid is now the couple mean distance to the fittest genotype and mean fitness. So, each time a better individual emerges, the context changes.

3.4   Experimental Studies

In this section, we present a set of results representing some typical examples of our experimentations. Each run, the GA works during 1000 generations, with 100 chromosomes each being 200 bits long. The classical mutation rate is fixed to 0.5 per chromosome (i.e. 0.0025 per bit) while the neutral mutation rate is set to 0.02. We choose to use the two points crossover at a rate of 0.7 and the 2-tournament selection scheme.

3.4.1   Order 1 Neutrality

The landscape designed for this experimentation is defined with a piecewise affine function in the unitation fitness plan. Its global outline looks like the letter V from which some segments (8 in number) are set to zero.

Without synthetic neutrality, the population is trapped in a local optimum (cf. figure 1). A contrario, with order 1 synthetic neutrality (MSN), after some accumulations followed by jumps through the landscape, the population succeeds to converge towards the global optimum (cf. figure 2).

Observing the mean fitness evolution, we can distinguish three steps before important fitness gains (cf. figure 3). These steps, and associated accumulation, corresponds to the exploitation times of neutral networks, while gains and jumps are linked to inter networks moves. We also notice a degradation of the mean fitness value before the gains and during the jumps. Indeed, during accumulations, the population explores the two subspaces 0Ω et 1Ω. This phenomenon increases the genotype polymorphism and produces the coming out of more diversified individuals, who are potentially degraded.

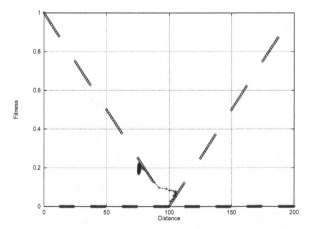

Fig. 1. Centroid trajectory without synthetic neutrality.

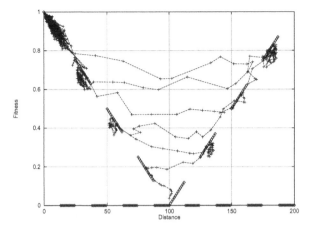

Fig. 2. Centroid trajectory with order 1 synthetic neutrality.

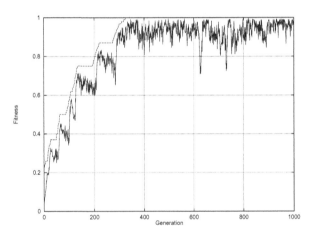

Fig. 3. Mean and max fitness evolution with order 1 synthetic neutrality.

Moves through neutral networks are equivalent to moves between the sub-spaces 0Ω and 1Ω. We can observe such moves on the figure 4. Note that this graph does not correspond to the same experiment than the previous one, since the length of the chromosome should be less than 64 bits. These migrations seem to be released by the exponential growth of particular genotypes.

The figure 5 shows the trajectory of the centroid obtained with the minimal synthetic neutrality when the crossover is turned off. The jumping behavior is still present and even more clearly observable. This indicates that the important feature in this experiments is the couple classical mutation and neutral muta-tion associated with the transformation of the structure of the search space.

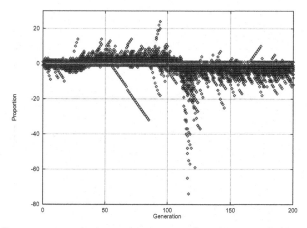

Fig. 4. Genotype cumulative proportions with order 1 synthetic neutrality.

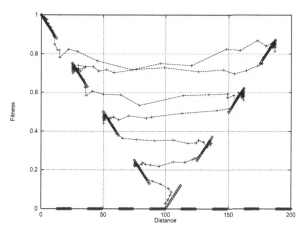

Fig. 5. Centroid trajectory with order 1 synthetic neutrality, no crossover.

Nevertheless, experiments on other problems show the importance of crossover to converge towards optimal solutions.

3.4.2   Order 2 Neutrality

The landscape designed for this experimentation was obtained with a blockwise function. The first block is evaluated by the function used previously with only 4 zero fitness segments. The second block is evaluated according to the number of bits equal to zero (inverted unitation).

Even with order 1 neutrality synthesis, the problem is still difficult and the population doesn't attain the global optimum (cf. figure 6). Studying the relative trajectory (see figure 7), we observe two distinct jumps which correspond to the coming out of genotypes that are fitter and very distant from the mean population. These jumps are induce by jumps between neutral networks.

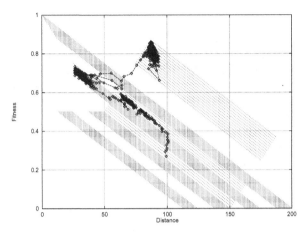

Fig. 6. Centroid trajectory with order 1 synthetic neutrality.

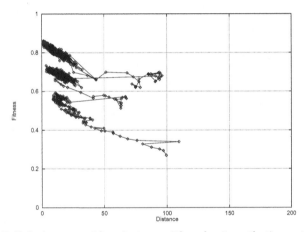

Fig. 7. Relative centroid trajectory with order 1 synthetic neutrality.

On the other hand, with order 2 synthetic neutrality, dynamics lead the population to follow a path that allows it to converge to the global optimum after two jumps (cf. figures 8 and 9).

As we see in figure 10, the mean distance of the first block goes like in the order 1 neutrality experimentation. While the mean distance of the second block evolves according to its inverted unitation. The two jumps of the centroid trajectory may be linked to jumps of the first block. This allows us to bring in light the existence of parallel dynamics that are probably responsible of the right convergence of the population.

Again, the population, when the crossover is switched off, exhibits the jumping behavior, as showed on figure 11.

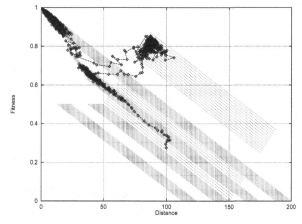

Fig. 8. Centroid trajectory with order 2 synthetic neutrality.

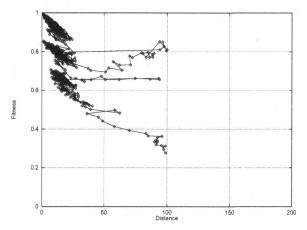

Fig. 9. Relative centroid trajectory with order 2 synthetic neutrality.

4   Conclusion

The main difference between neutral and selectionist theory of evolution relies on the explanation of the natural polymorphism. Indeed, genetic diversity is undoubtedly useful for the evolution, but this diversity is paradoxical since the selective pressure leads the populations to become homogeneous. Some people bring up selective arguments, others believe in neutral mutation hypothesis. The truth should lay between them.

This discussion can take place in GA field which is exposed to premature convergence, that is the loss of diversity. Until now, the classical techniques were selectionist, as sharing, which, modifying the selective value, allows to escape from local optima. We showed that synthetic neutrality with *ad hoc* evaluation functions also allows to escape from such optima. Neutral networks with a particular structure can be exploited by the algorithm.

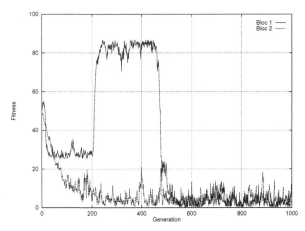

Fig. 10. Time evolution of the mean distance per block with order 2 synthetic neutrality.

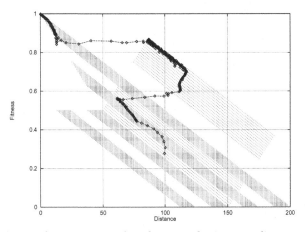

Fig. 11. Centroid trajectory with order 2 synthetic neutrality, no crossover.

Synthetic neutrality is for neutral theory of evolution as sharing for selectionist theory. In Nature, there is no clear evidence to use the neutral or the selectionist way of explanation for polymorphism. In the same way, in the artificial evolution field, the joint use of sharing and synthetic neutrality may and should be considered.

The neutrality concept has allowed us to review and to extend the concept of duality ([3,4]). This has led us to build new analyze oriented tools for the dynamics; thus, we have shown the exploitation of neutral networks by a GA. Further works will deal with the importance of the synthetic neutrality compared to the inherent neutrality of an adaptive landscape. Indeed, we expect to be able to dissociate the different dynamics due to the inherent neutrality, the synthetic neutrality and the selective pressure. Of course, we also have to study the capacity of the synthetic model to solve more realistic problems. Finally, it would be very interesting to more deeply analyze the part played by crossover in the diffusion mechanisms.

References

1. L. Barnett. Evolutionary dynamics on fitness landscapes with neutrality. Master's thesis, University of East Sussex, Brighton, 1997.
2. C. L. Barrett, H. Mortveit, and C. M. Reidys. Elements of a theory of simulation ii : sequential dynamical systems. *Appl. Math. & Comp.*, 1998.
3. P. Collard and J.P. Aurand. DGA: An efficient genetic algorithm. In A.G. Cohn, editor, *ECAI'94: European Conference on Artificial Intelligence*, pages 487–491. John Wiley & Sons, 1994.
4. P. Collard and C. Escazut. Relational schemata: A way to improve the expressiveness of classifiers. In L. Eshelman, editor, *ICGA'95: Proceedings of the Sixth International Conference on Genetic Algorithms*, pages 397–404, San Francisco, CA, 1995. Morgan Kaufmann.
5. T.D. Collins. Using Software Visualization Technologies to Help Evolutionary Algorithm Users Validate their Solutions. In Thomas Bäck, editor, *Proceedings of the seventh International Conference on Genetic Algorithms*, pages 307–314. Morgan Kaufmann, 1998.
6. C. V. Forst, C. Reidys, and J. Weber. Evolutionary dynamics and optimization: Neutral networks as model-landscapes for rna secondary-structure folding-landscapes. In *Advances in Artificial Life*, volume 929 of *LNAI*. Springer Verlag, 1995.
7. W. Hordijk. Correlation analysis of the synchronizing-ca landscape. *Physica D*, 107:255–264, 1997.
8. M. A. Huynen, P. F. Stadler, and W. Fontana. Smoothness within ruggedness: the role of neutrality in adaptation. In *Proc. Natl. Acad. Sci. (USA)*, volume 93, pages 397–401, 1996.
9. M. Kimura. *The Neutral Theory of Molecular Evolution*. Cambridge University Press, Cambridge, UK, 1983.
10. S. Wright. The theory of gene frequency. In *Evolution and the genetics of Population*, volume 2, pages 120–143. University of Chicago Press, 1969.

Two Evolutionary Approaches to Design Phase Plate for Tailoring Focal-Plane Irradiance Profile

Sana Ben Hamida, Alain Racine, and Marc Schoenauer

Centre de Mathématiques Appliquées
Ecole Polytechnique
91128 PALAISEAU, FRANCE

Abstract. The goal is to design the 2-dimensional profile of an optical lens in order to control focal-plane irradiance of some laser beam. The numerical simulation of the irradiance of the beam through the lens, including some technological constraints on the correlation radius of the phase of the lens, involves two FFT computations, whose computational cost heavily depends on the chosen discretization.

A straightforward representation of a solution is that of a matrix of thicknesses, based on a $N \times N$ (with $N = 2^p$) discretization of the lens. However, even though some technical simplifications allow us to reduce the size of that search space, its complexity increases quadratically with N, making physically realistic cases (e.g. $N \geq 256$) almost untractable (more than 2000 variables). An alternative representation is brought by GP parse trees, searching in some functional space: the genotype does not depend any more on the chosen discretization.

The implementation of both parametric representation (using ES algorithms) and functional approach (using "standard" GP) for the lens design problem are described. Both achieve good results compared to the state-of-the-art methods for small to medium values of the discretization parameter N (up to 256). Moreover, preliminary comparative results are presented between the two representations, and some counter-intuitive results are discussed.

1 Introduction

Many optimization problems actually look for a function: unknown profiles can be seen as spatial functions, unknown commands can be described as time-dependent functions, ... Two approaches are then possible: standard numerical methods usually discretise the domain of definition of the unknown function, and transform the search space into a space of vectors of parameters (one value per discrete point). The original optimization problem is thus amenable to parametric optimization. But the accuracy of the solution is then highly dependent on the discretization, and the finer the discretization, the larger the parametric search space. On the other hand, Evolutionary computation offers an alternate approach where the search space is some functional space, independent of any discretization.

C. Fonlupt et al. (Eds.): AE'99, LNCS 1829, pp. 266–276, 2000.

This paper presents a case study of such situation. The goal is to design a lens in order to control focal-plane irradiance of some laser beam. Irradiance control allows to concentrate laser beam energy on a particular region of the focal-plane of a focusing lens. Physicists use this technology to experiment a specific kind of nuclear fusion : *Inertial Confinement Fusion*. A tiny pellet of fuel is heated to a very high temperature by powerful beams of energy. Thus the laser inertial confinement process produces powerful bursts of fusion energy.

The best-to-date method to optimize the profile of the phase plate is simulated annealing [YLL96]. Nevertheless, this approach requires a significant computational time (120h of CPU time on a Cray YMP-2 computer). Other heuristic methods [SNDP96,fLEa,fLEb] can be used to provide an approximate phase profile. But these lenses are not efficient, i.e. they induce a drastical loss of energy outside the target: the spared computational time is balanced by a loss of precision.

We propose in this paper to use the *Evolutionary Computation* paradigm to handle the lens design problem. The choice of a representation is well known to be a crucial step in any Evolutionary Algorithm (EAs) – see e.g. the early debate on binary vs real encoding for real-valued parameters. On the other hand, EAs are flexible enough to be able to handle non-standard search spaces, such as spaces of variable length lists, graphs, etc. Along those lines, many representations have been designed to handle functions, among which Genetic Programming has proven very efficient on a number of problems (see e.g. [SSJ+96,Koz94] among others).

In the case of the lens profile design, the computation of the fitness of a given profile requires some numerical simulation of optical propagation of a laser beam, and the best-to-date methods for that involve Fast Fourier Transforms (FFTs), based on some $N \times N$ discretization of the profile (see section 2). Hence a "natural" representation for a profile is the 2-dimensional matrix that will be used to compute those FFTs. But, even though working in the time domain allows us here to decrease the number of unknown parameters (see again section 2 for the details), the size of this parametric search space increases quadratically with the chosen discretization size N. So even without considering the computational cost of a single fitness evaluation, it is commonly acknowledged that the number of generations before convergence of any EA increases at least linearly with the size of vector genotype, at least in the binary use (see e.g. [TG93,Cer96]).

This is where functional approaches, among which Genetic Programming (GP), can take over: indeed, if the lens profile at a given point is represented as a function of that point, the size of the search space is independent of any discretization. Hence the number of generations to reach a given accuracy will hopefully be independent of any discretization, too – even though the computational cost of a single fitness evaluation will still depend on the discretization (i.e. the number of points where GP trees need to be evaluated still increases quadratically with the discretization parameter N).

Both approaches are presented in this paper: the parametric representation is handled by a $(\mu + \lambda)$ Evolution Strategy, and the non-parametric functional approach uses "standard" Genetic Programming to represent solutions by parse trees.

This paper first describes the application background and presents some feasibility constraints (section 2). The parametric representation, together with the associated ES algorithm, is introduced in section 3.1, while the functional approach based on Genetic Programming is detailed in section 3.2. The results of both algorithms are presented and compared in section 4. Both approaches give very good results compared to the state-of-the-art methods. However,the comparaison of both approacheswhen N increases yields some counter-intuitive results.

2   Focal-Plane Irradiance Problem

Consider a laser beam, specified by its scalar electrostatic field. The problem consists in designing a continuous distributed phase plate (the unknown lens shape) to produce a specified irradiance profile at the focal-plane of a given focusing lens.

Without any phase plate, the focal-plane profile corresponds to a single (very energetic!) peak at the focal point. In order to achieve the *inertial confinement fusion*, it is mandatory to be able to illuminate a given small circular target around the focal point of the focusing lens. Hence a phase plate is added before the focusing lens and its shape directly acts on the irradiance profile.

More precisely, the optical system (See Figure 1) is composed by :

– a large pupil to control the complete illumination of the phase plate.
– a phase plate (to be determined) to modify the distribution of the phase along the section of the laser beam.
– a focusing lens
– a *virtual screen* at the focal-plane to measure the focal profile.

2.1   Analytic Formulation

Figure 1 shows the experimental layout. First, the *pupil* produces a scalar electrostatic field:

$$E_0(x,y) = e^{-\pi \left(\frac{\sqrt{(x^2+y^2)}}{R} \right)^8}$$

The new field $E_1(x,y)$ generated by the phase plate $\varphi(x,y)$ is:

$$E_1(x,y) = E_0(x,y).e^{i\varphi(x,y)}$$

Thus, the field at the focal-plane is obtained by a simple Fourier Transform:
$E_2(x,y) = \mathrm{FT}(E_1(x,y))$

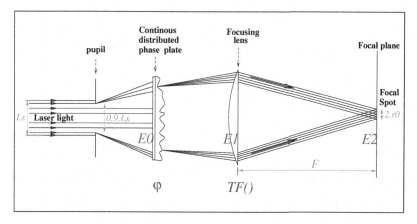

Fig. 1. The optical system. All physical values have been normalized.

Finally, the *relative* lighting intensity at the focal-plane is given by the intensity of the filed E_2, the Fourier transform of E_1:

$$I(x,y) = (E_2(x,y))^2 = (\text{FT}(E_1(x,y)))^2$$

2.2 The Optimization Problem

The goal is to design a phase plate such that a small circular target of radius r_0 on the focal plane is uniformly illuminated by the laser beam (see Figure 2-a): the perfect solution shows a radial step-like profile on the focal plane – which will be smoothed to a super-Gaussian profile of order 8 (i.e. $t(r) = \exp(-(r/r_0)^8)$ in the following.

The radial profile $\overline{I}(r)$ from $I(x,y) \equiv I(r,\theta)$ is given by:

$$\overline{I}(r) = \frac{1}{2\pi r} \int_0^{2\pi} I(r,\theta)\, d\theta$$

The goal is to design a phase-plate whose illumination profile on the focal plane will be as close as possible from the super-Gaussian profile (see Figure 2-b). The cost function (to be minimized) $\mathcal{E}rr$ becomes (after renormalization of $\overline{I}(r) \to \widetilde{I}(r)$) :

$$\mathcal{E}rr = \int_0^{\infty} \left[t(r) - \widetilde{I}(r) \right]^2 dr \tag{1}$$

2.3 Manufacturing Constraints

Some technological constraints have to be taken into account: manufacturers are not able to make very rough phase-plates. To ensure an acceptable smoothness

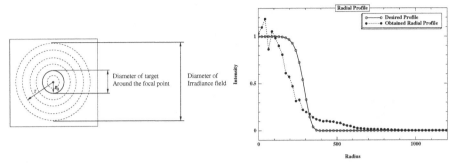

(a) The target in the focal plane.   (b) Desired and sample radial profiles.

Fig. 2. The optimization problem: fitting the desired radial profile.

of all designs, a common practice is to filter out the high frequencies of the phase-plate in the frequency domain (after applying Fourier transform), before transforming the filtered signal back into the space domain.

This leads to consider a new unknown field $Z(x, y)$, and to compute the phase plate φ by filtering out the high frequencies in Z as shown in Table 1 below.

Table 1. Smoothing the phase plate: the filter simply zeroes out high frequencies

$$Z(x, y) \underset{FFT}{\longrightarrow} \tilde{Z} \underset{Filtering}{\longrightarrow} \tilde{Z} \times Filter \underset{FFT^{-1}}{\longrightarrow} \varphi(x, y)$$

2.4   Working in Fourier Space

Table 1 suggested a simplification of the search space: instead of working on Z, it is possible to consider directly the filtered Fourier transform $\tilde{Z} \times Filter$ as the unknown field. The number of non-zero terms in $\tilde{Z} \times Filter$ is much smaller than the number of terms in Z as all high frequencies correspond to a zero in the filter. Whereas it is clear that this will be an advantage for the parametric representation (see section 3.1 below), it nevertheless decreases the number of points where the unknown field $\tilde{Z} \times Filter$ has to be computed in the functional approach (section 3.2).

The number of non-zero values in $\tilde{Z} \times Filter$ is given in Table 2 for the different values of the discretization N used in the different FFT involved in the fitness.

Table 2. Non-zero terms in $\tilde{Z} \times Filter$ for different values of the discretization parameter N

$N = 32 \longrightarrow nval = 32$
$N = 64 \longrightarrow nval = 148$
$N = 128 \longrightarrow nval = 560$
$N = 256 \longrightarrow nval = 2284$

2.5   Fitness Function

From the above consideration, the fitness function used thereafter can be graphed as in Figure 3, the comparison between both profiles being given by equation (1).

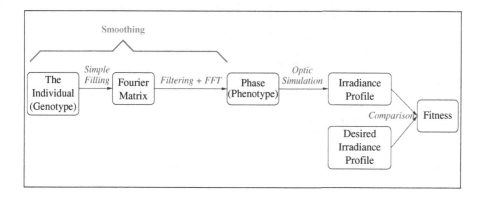

Fig. 3. Fitness computation

The remaining open issue is the representation of the non-zero field $\tilde{Z} \times Filter$ - which will be thoroughly addressed in next section.

3   The Evolutionary Algorithms

3.1   Parametric Representation

The most "natural" representation for the field $\tilde{Z} \times Filter$ of Table 1 is the direct encoding of all non-zero terms into a real-valued matrix - the size of the matrix is a function of the discretization parameter N as given in Table 2 (though it is not N^2, it is quadratically increasing with N).

We are then back into the familiar framework of parametric optimization, and have chosen a self-adaptive Evolution Strategy algorithm to address that problem (see [Sch81,B95,BS93,BHS97] for more details on standard ES with self-adaptive mutation parameters).

Briefly, a standard deviation is attached to each design variable, and the mutation operator first mutate the standard deviation following a log-normal mutation before modifying the design variable with a Gaussian noise using the new value of its standard deviation.

The recombination used is the standard ES "global discrete recombination" for both the standard deviations and the design variables. Note that a 2-dimensional crossover operator was also tried (the unknown parameters are indeed a 2-D matrix [KS95]) which did not give better results.

The first population is initialized randomly in the interval $[-1500, 1500]$. It is interesting to note that initial results with a much smaller initialization interval $([-1, 1])$ showed much slower convergence of ES ([RHS99]).

All parameters are given in Table 3.

Table 3. ES parameters

Population size: μ	50
Number of offspring: λ	100
Selection Mode	$(\mu + \lambda)$
Mutation parameters	
Initial std. dev.: σ_0	0.3
Global update: τ [1]	1
Local update: τ_L [1]	0.7
Initialization	$[-1500, 1500]$

[1]These parameters are modified according to the number $nval$ of parameters to optimize (see [BS93]).

3.2   Functional Representation: GP

As discussed in the introduction, Evolutionary Algorithms offer alternate representations for functions – and the idea here is to represent directly the unknown field $\tilde{Z} \times Filter$ as a parse tree, using "standard" Genetic Programming as the optimization algorithm [Koz92,Koz94]. Table 4 is the standard GP tableau, giving all parameters.

Table 4. Tableau for the GP algorithm

Set of nodes	$\{+, -, *, sin\}$
Set of terminals	$\{x, y, \mathbb{R}\}$
Window of discretization	$[-32, 31]^2$
Raw and stand. fitness	**Given by Equ. (1)**
Wrapper	**See Table 1**
Population size	**100**
Selection Mode	**Tournament (Size=10)**
Replacement	**Generational with elitism**
Crossover probability	**0.6**
Mutation probability	**0.4**
Types of mutations	
	Promotion of a branch
	Random Replacement of a branch
	Node and terminal permutation
	realvalued terminals Gaussian mutation
A combination of **3** mutation operators is used for each mutation	
Maximal depth of an individual	**14**
Minimal depth at initialization	**3**
Interval for initialization of realvalued terminals	[-100,100]
Std. dev. for Gaussian mutation of realvalued terminals	**10**

4 Results

Figure 4 is a graphical view of the average and best fitnesses (over 11 runs) for both approaches.

The first important result is that the results are very good indeed: an error of 0.015 is reached in 5000 generations –for the 128 × 128 case– corresponding to a total computing time of about 10 hours of computing time for a Pentium II 350 Mhz, to be compared to the 120 hours of Cray YMP-2 for the 64 × 64 discretization. The best solution is given in Figure 5, showing a very good fitting of the target irradiance profile.

But when it comes to compare the parametric and the functional approaches, some surprising facts arise.

In the early generations, both ES and GP reach quickly reasonable performance, thanks to the large initialization interval for ES and to the genetic operators for GP, that can be considered at this stage as exploration operators. Indeed, GP-mutations and GP-crossover allow the algorithm to make large jumps in the search space since a small modification of a tree generates a drastically different behavior. These results tallies with Angeline's work concerning the link between crossover and macro-mutation [Ang97].

But, in a second part, while ES still improving the best fitness, GP seems to have much difficulty to refine the solutions and reach smaller errors.

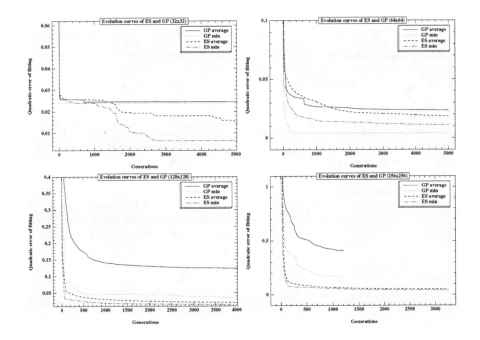

Fig. 4. The average (out of 11 runs) and best fitness plots for both GP and ES for N=32, 64, 128 and 256.

With ES, even though the convergence remains very slow, the fitness steadily decreases –thanks to the self-adaptation process– and this is true up to 15000 generations for each discretization level. With GP, the algorithm roughs out very efficiently the exploration of the fitness landscape, but is unable to refine the solution. Indeed GP operators lack precision and are not able to precisely fine-tune the phenotypic behavior from the genotypic representation. A possible improvement would be to also use self-adaptation to tune the realvalued terminals in GP rather than performing Gaussian mutation with fixed standard deviation – future work will investigate this possibility.

Another point that enforces this explanation is that the best results for GP are provided with a rather large tournament size (10 out of a population of 100). Indeed, this forces to concentrate the GP exploration around the best solution to improve the refinement process. However this approach rapidly reduces the diversity in the population, and GP is then searching in a quite narrow region of the search space. A way to increase diversity could be to use much large population sizes. On-going work focuses on that issue.

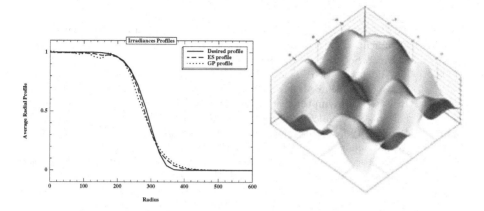

Fig. 5. The best radial profiles (ES & GP) and the best phase plate (ES) for $N = 128$

5 Conclusion

In this paper, a difficult real-world problem has been tackled by Evolutionary Computation using two different representations: the straightforward parametric representation by a matrix of real numbers, optimized using an ES+ algorithm, and the functional representation allowed by Genetic Programming.

Both algorithms gave good results compared to the state-of-the-art results, but the comparison of the parametric and the functional approaches lead to some surprising results. Whereas the GP functional approach was expected to be less sensitive than the ES parametric approach, the latter was able to obtain better-tuned results – whatever the discretization parameter. Our explanation lies in the weakness of our GP implementation to fine-tune the realvalued terminals – and hence locally optimize the good solutions it has found.

But the ultimate functional approach would be to use even further the analytic form of GP trees, postponing the numerical discretization as much as possible: We are presently working on a a pure symbolic system based on GP – using symbolic maths to compute all Fourier Transform whenever possible. This approach should provide a functional representation of the optimized phase plate instead of a discrete representation, hopefully avoiding most side effects generated by the discretization.

Acknowledgements. We want to thank Antoine Ducoulombier for pointing out the right initialization interval.

References

Ang97.   Peter J. Angeline. Subtree crossover: Building block engine or macromutation? In J. R. Koza and al., editors, *GP97: Proceedings of the $2^n d$ Annual Conf.* Morgan Kaufmann, 13-16 July 1997.

B95.   T. Bäck. *Evolutionary Algorithms in theory and practice*. New-York:Oxford University Press, 1995.

BHS97.   Th. Bäck, U. Hammel, , and H.-P. Schwefel. Evolutionary computation: Comments on the history and current state. *Transactions on Evolutionary Computation*, 1(1):3–17, 1997.

BS93.   Th. Bäck and H.-P. Schwefel. An overview of evolutionary algorithms for parameter optimization. *Evolutionary Computation*, 1(1):1–23, 1993.

Cer96.   R. Cerf. An asymptotic theory of genetic algorithms. In J.-M. Alliot and al., editors, *Artificial Evolution*, volume 1063 of *LNCS*. Springer Verlag, 1996.

fLEa.   Laboratory for Laser Energetics. Distributed phase plates for super gaussian focal-plane irradiance profiles. *LLE Review*.

fLEb.   Laboratory for Laser Energetics. High-efficiency distributed phase plate generation and characterization. *LLE Review*.

Koz92.   J. R. Koza. *Genetic Programming: On the Programming of Computers by means of Natural Evolution*. MIT Press, Massachussetts, 1992.

Koz94.   J. R. Koza. *Genetic Programming II: Automatic Discovery of Reusable Programs*. MIT Press, Massachussetts, 1994.

KS95.   C. Kane and M. Schoenauer. Genetic operators for two-dimensional shape optimization. In J.-M. Alliot and al., editors, *Artificial Evolution*, number 1063 in LNCS. Springer Verlag, Septembre 1995.

RHS99.   A. Racine, S. Ben Hamida, and M. Schoenauer. Parametric coding vs genetic programming: A case study. In *Second European Workshop on Genetic Programming*, 1999.

Sch81.   H.-P. Schwefel. *Numerical Optimization of Computer Models*. John Wiley & Sons, New-York, 1981. 1995 – 2nd edition.

SNDP96.   M. D. Perry S. N. Dixit, M. D. Feit and H. T. Powell. Designing fully continuous phase screens for tailoring focal-plane irradiance profiles. *Optics Letters 21*, 1996.

SSJ+96.   M. Schoenauer, M. Sebag, F. Jouve, B. Lamy, and H. Maitournam. Evolutionary identification of macro-mechanical models. In P. J. Angeline and Jr K. E. Kinnear, editors, *Advances in GP II*, Cambridge, MA, 1996. MIT Press.

TG93.   D. Thierens and D.E. Goldberg. Mixing in ga. In S. Forrest, editor, *Proceedings of the 5th International Conference on Genetic Algorithms*. Morgan Kaufmann, 1993.

YLL96.   T. J. Kessler Y. Lin and G. N. Lawrence. Design of continuous surface-relief phase plates by surface-based simulated annealing to achieve control of focal-plane irradiance. *Optics Letters*, 1996.

A Shepherd and a Sheepdog to Guide Evolutionary Computation?

Denis Robilliard and Cyril Fonlupt

Laboratoire d'Informatique du Littoral
Université du Littoral
BP 719
62228 CALAIS Cedex
FRANCE
phone: +33-3-21-97-00-46
fax: +33-3-21-19-06-61
email: {robillia,fonlupt}@lil.univ-littoral.fr

Abstract. Memory is a key word in most evolutionary approaches. One trend in this field is illustrated by the PBIL method that stores statistical information on the values taken by genes of best individuals. The model we are presenting follows these lines, and stores information both from good individuals (*attractor* memory) and bad ones (*repoussoir* memory). We show the interest of this model on classical binary test instances. Next we propose to test variants of this strategy for solving a non binary optimization problem: the traveling salesperson problem (TSP). We discuss the difficulties that one must face to tackle a non binary representation, and present a solution adapted to the TSP. We put into evidence the lack of significant differences between results from these strategies, and argue about some characteristics of the TSP search space that could explain this behaviour.

1   Introduction

Looking at basics in Darwinian evolution, we can identify three concepts: the selection (natural selection but surely also plain random, in the form of genetic drift for example), the evolution material (genes and individuals), and the evolution opportunities, that is the ways the evolution material can be subject to alteration (mutation, cross over, ...). Transposing this into artificial evolution, we want to focus on some aspects of the last two concepts, evolution material and evolution opportunities, and notably on how they are expressed and used in the field of optimisation.

In Evolutionary Computation, the evolution material is usually represented by encoded genotypes, that constitute the individuals forming the population in population-based methods. Any individual, storing its genotypic information, can thus be seen as a memory. But it is interesting to see that memory may take different forms in various algorithms such as hill-climbing strategies, population based incremental strategies, $(\lambda + \mu)$ evolution strategies, and more complex methods like Genetic Algorithms, Tabu search or Simulated Annealing...

C. Fonlupt et al. (Eds.): AE'99, LNCS 1829, pp. 277–291, 2000.

All these search schemes, either simple or complex, make use of shared, or at least partly shared, memory, as a basic method for keeping and transmitting information, notably to offspring. It is indeed this paradigm that allows us to differentiate them from true random search. Anyway, this memory does not need to assume the same form as can be seen in the following examples:

- in the case of the hill-climbing strategies (e.g. [1,2]), the memory is the individual itself. A small perturbation is applied to it and the resulting offspring is kept if it improves its parent.
- in the case of a more complex scheme, e.g. the Tabu search [3], the memory is explicitly described as a Tabu list. But new individuals are usually created by adding perturbation to old ones, so, here again, part of the memory is the individual itself;
- in the case of Population-Based Incremental Learning (PBIL) [4,5,6], a summary is kept of the mean values taken by genes of the best individuals at every generation. The next generation will be created from scratch, using this information as a probability distribution. Notice that individuals are not directly used as memory;
- the $(\lambda + \mu)$ evolution strategies can be viewed somewhat as an extension of the basic hill-climbing strategies, where λ stands for the number of hill-climbers and μ for the number of "concurrent" trials for the hill-climbers. Memory is again solely in the individuals.

The second concept we are interested in, is the way new individuals are generated. We can see from the previous examples of algorithms that small perturbations, ressembling to slight natural mutations, are often used, notably in local search approaches (hill-climber, Tabu). Many cross-over operators have also been devised for sharing information in genetic algorithm. However allowing large mutations may also be used, as was demonstrated by the PBIL scheme. But the most striking difference with nature, is that mutation is not always blind:

- it may be blind, as in the canonic genetic algorithm;
- it can be guided towards systematic individual improvement, as in the hill climber strategy;
- it can be oriented to prevent a new individual to be too close to its parent in genotypic space, as in the Tabu search;
- it can be oriented to prevent a new individual to be both too close or too far to its "parents" in genotypic space, as in the Scatter Search method;
- it can enhance the likeness of new individuals to already encountered good solutions, as in the PBIL scheme;
- it can also enhance the difference between already seen "bad" solutions and the newly generated ones, as was proposed in the *repoussoir*[1] scheme by Sebag and Schoenauer [7].

---

[1] "Repoussoir" comes from the french word to call a person you do not want to be or look like.

This paper studies how memory of previously encountered individuals, be they either good or bad solutions at the time we first met them, may be used as a guideline for generating new genotypes, thus providing the evolution material and evolution opportunities. As we said above, there were some decisively influential papers that inspired us for this work, notably the PBIL paper from Baluja [4,5,6], and the "repoussoir" studies from Schoenauer [7], and Seebad, Schoenauer and Peyral [8].

Our approach is based on a $(\lambda + \mu)$ evolution strategy. We keep a memory of the best encountered solutions, what we call an *attractor*, and also a memory of the worst encountered solutions, in a *repoussoir*. Every individual in the population is subject to a mutation phase that is organized either along the attractor strategy or the repoussoir strategy. In the attractor strategy, the individual mutates some of its genes towards the allele values that are present in the attractor memory. In the repoussoir case, the individual mutates some of its genes towards values that differ from those stored in the repoussoir. When an individual engages in a strategy, it keeps on with it until he fails to produce offspring better than itself for a given number of generations. This may be viewed as if we were trapped in a "local optimum" in the local search paradigm. In this case, we swap to the other strategy, to try to escape from this local optimum. In a few words, we either push individuals towards good solutions or push them away from bad ones, hence the title "Shepherd" (attractor) and "Sheepdog" (repoussoir). This behaviour of swapping strategies to escape locked situation is shown to have beneficial effects when solving classical binary problems. This is the first extension that we bring to previous works.

The other originality of this study is the application of a set of $(\lambda + \mu)$ evolution schemes, using different repoussoir-like memory models, to a non binary optimization case, namely the traveling salesperson problem (TSP). In the binary case, it is rather easy to mutate a gene towards or away a given value: you just copy it or invert it. In the TSP case it is no more straightforward, and we propose mutation operators that preserve the semantic of the model. We are also faced with storage size constraints for the attractor and repoussoir memory, that are solved using a replacement scheme, where new information stochastically replaces older one. The results from these various strategies are compared. The lack of significant differences when using such memory model is then discussed in the light of the complexity of the TSP search space.

The paper is structured as follows. In section 2 we outline the "shepherd and sheepdog" algorithm. We present some results and some comparisons with other schemes in the case of binary search space problems in section 3. In section 4, we introduce the way to adapt attractor/repoussoir approach to the non-binary search space of the TSP and we discuss experimental results in section 4.3. Directions for future works and conclusions are presented in section 5.

2   The Guided Hill-Climber

In this section we describe three algorithms based on a $(\lambda + \mu)$ evolution strategy model. Each of these algorithms implements aspects of the Attractor/Repoussoir paradigm, in the case of binary representation search space. We compare results on binary optimization problems with those published in [7].

2.1   The Attractor Algorithm

The attractor algorithm was proposed in [8]. It can be described as a variant of the PBIL technique. In both algorithms, an array of floatting point values stores a kind of "average" of the genes values of the B best solutions found at each generation of the run. Parameter B defines the number of elite individuals that we want to use for updating the attractor memory, typical values ranging from 1 (the best member only) to $1/10^{th}$ of the population size . In every generation, there is a relaxation phase, in order to give older information less importance than what is given to the genes values of the current best individuals. This is indeed akin to the "evaporation phase" that takes place before pheromon update in Ant Colony Systems [9,10]. We call this memory of best genes values the attractor. In the PBIL case, at every generation the population is created from scratch, using only the probabilistic information from the attractor. In our algorithm, instead of creating the population from scratch, we use a hill-climber approach, perturbing the current generation individuals to obtain the next ones, and using the attractor memory of previous best individuals to decide of the mutation points. So we draw information both from the attractor and directly from individuals of the previous generation. The bits to mutate are drawn by a tournament method, advantaging bits that are most dissimilar from the corresponding ones in the attractor memory. These bits are inverted, so, in one word, this algorithm tries to make individuals come nearer to the attractor. Notice that we don't copy best bits from the attractor, but rather invert bits ressembling the least to those in the attractor memory: this choice has been made to limit premature convergence.

2.2   The Repoussoir Algorithm

The repoussoir algorithm was proposed in [7]. It is designed along the same general lines as the attractor. The past unfruitful trials are stored as an "average" individual called repoussoir. We use again a hill-climber approach, perturbing the current generation individuals to obtain the next ones, and using the repoussoir to decide of the mutation points. The bits to mutate are drawn by a tournament method, advantaging bits that are most similar to the corresponding ones in the repoussoir. These bits are inverted, so, in one word, the search algorithm tends to push individuals away from the repoussoir.

2.3   The Shepherd and Sheepdog Algorithm

In the shepherd and sheepdog paradigm, we propose to use a mixed approach. Instead of basing the hill-climbers walks on one strategy, be it attractor or repoussoir, we swap from one to the other when the algorithm is unable to improve individuals for a given number of generations denoted by parameter G. Such locked situations, where improvement is difficult to obtain, may be considered as equivalent to local optima in the framework of local search. Changing the strategy implies changing the fitness landscape and thus gives a chance to escape from the trap.

The shepherd and sheepdog scheme can also be described by the metaphor of the magnet: sometimes we use a magnet to attract solutions, but sometimes we also use another magnet, with inverse polarity, to push them forward.

The pseudo-code of attractor, repoussoir and "shepherd and sheepdog" techniques is given in Appendix.

3   Test Functions in Binary Search Spaces

For testing the approach introduced in Section 2, we use the functions introduced in [7] as well as a function known to be difficult presented in [11]. In this study of binary search space problems, the focus is on finding if the shepherd and sheepdog approach can improve the performance of an attractor only or repoussoir only scheme. Results are also compared with two standard hill-climbing methods [5], and a set of search methods from [7]. It is important to notice that all algorithms use the same number of calls to the evaluation function, this number being taken from [7]. This implies that the shepherd and sheepdog must take advantage of a cooperation between both method if it is to obtain an advantage against attractor only or repoussoir only.

3.1   Test Functions

The first three functions denoted as f_1, f_2 and f_3 were introduced in Baluja's work [4] as a benchmark for several evolutionary algorithms. They are coded on 900 bits, each variable being coded on 9 bits ranging between $[-2.56, 2.56]$. Function f_4 introduced in [11] is a 10-dimensional function, it involves 10 numerical variables being coded on 10 bits. The range of each variable lies in $[-20, 20]$.

$$y_1 = x_1$$
$$y_i = x_i + y_{i-1}, i \geq 2 \qquad\qquad f_1 = \frac{100}{10^{-5}+\sum_i |y_i|}$$

$$y_1 = x_1$$
$$y_i = x_i + \sin(y_{i-1}), i \geq 2 \qquad\qquad f_2 = \frac{100}{10^{-5}+\sum_i |y_i|}$$

$$f_3 = \frac{100}{10^{-5}+\sum_i |.024 \times (i+1)-x_i|}$$

$$f_4 = \sum_{i=1}^{10} \sin \sqrt{|40x_i|} + \frac{20-x_i}{20} \quad -20 \leq x_i \leq 20$$

3.2 Conclusions

All algorithms are allowed 200000 functions evaluations in order to allow a fair comparison. The experiments were repeated twenty times per test problem, we set the same parameters than [7] ($\mu = 10$, $\lambda = 50$, $R = 50\%$, $T = 30$ and $\alpha = .01$). "Repoussoir only" results are in line "REP", "Attractor only" in column "ATTR", adn "Shepherd and Sheepdog" in column "SHEP".

Table 1. Results on binary space test functions.

ALGO	TEST FUNCTIONS			
	f1	f2	f3	f4
HC1	1.04	3.08	8.07	18.68
HC2	1.01	3.06	8.10	n.a.
SGA	1.96	3.58	9.17	n.a.
GA-scale	1.72	3.68	12.3	n.a.
AES	2.37	3.94	9.06	n.a.
TES	1.87	3.61	10.46	n.a.
PBIL	2.12	4.40	**16.43**	n.a.
NS	2.81	4.24	12.8	n.a.
REP	2.51	*3.49*	14.12	19.63
SHEP	**2.89**	**3.59**	**14.3**	**19.73**
ATTR	2.82	3.34	11.07	19.62

Although results are not always strinkingly better, we can see from Table 1 that the shepherd and sheepdog is able to improve performances, keeping in mind that the number of function evaluations is the same in every cases[2]. Only the PBIL method performs better on one of the test functions, with the shepherd and sheepdog getting the second place in front of the repoussoir and attractor schemes.

One drawback is this algorithm is that it seems very sensitive to two parameters: M the number of bits to mutate when perturbing an individual and G the number of generations without improvement before swapping strategy. A future step would be to dynamically adapt the value of M and G during the search. It could also be interesting to determine off-line the value of M by computing epistasis measures on the problem. Such measures could provide detection of the bits that are most sensitive to mutation, as we have shown in [12].

---

[2] When solving function f_2 with the "repoussoir only" scheme we have been unable to find again the result given in [7]; we give the value computed with our implementation (italic font).

4   A Case of Non Binary Representation: The TSP

The traveling salesperson problem (TSP) is a classic of combinatorial optimization (see for example [13] as a reference book). It is given as a set of N cities coordinates (or equivalently the inter-cities distance matrix), and the objective is to find a tour as short as possible, visiting every cities. It has been studied intensively, notably in the EA community. Evolutionary inspired methods seem to be among the best heuristic schemes nowadays. Algorithms like Inver-over Operator [14], the Repair and Brood Selection [15], Ant Colony Systems [16,10], and the Edge Assembly Crossover [17,18] are certainly among the most powerful heuristics for this problem.

4.1   Problems Arising from Non Binary Representation

The case of non binary search space is a tricky problem for designing an attractor/repoussoir algorithm. We face several problems:

- how to define an attractor or repoussoir memory adapted to non binary representation ? In the TSP case, storing every edges that are encountered leads us theoretically to a N^2 storage size. Using information from such a big storage is uneasy, since most of it will be useless in the last stages of the run: tournaments should imply too many members in order to have a chance of selecting sparse good genes;
- it is a known fact in planar TSP studies that the most promising area of the search space is a small compact subset, sometimes called "massif central" or "big valley" [19,20]. Once we are in the massif central, isn't the repoussoir technique at risk to push individuals away from it? Improving and keeping solutions in this promiseful zone is the hard part of any good planar TSP algorithm;
- furthermore, in the case of the binary search space, copying or inverting bits is an easy way to get closer or farther any given individual. In the TSP case, there is no such thing as inverting an edge, and copying an edge does usually lead to non hamiltonian graphs that need to be repaired to form a valid tour.

In the following section, we provide some clues on how to extend the memory model for the planar TSP. We document our implementation scheme and we provide some experimental results on well known TSP instances[3] from the TSPLIB [21].

4.2   Implementing Attractor and Repoussoir Scheme for the TSP

In order to keep the semantics of memory that was presented in the binary space case, we propose a model where edges linking two cities correspond to bits.

---

[3] the number used as suffix in TSPLIB instance names is the number of cities in the instance.

- The attractor (resp. repoussoir) is constructed as a list of fixed size $4 * N$, where N is the instance size. An entry in the list is a two valued structure that stores the edge together with its attraction or repulsion value. The higher this value, the more attractive or repulsive the edge is;
- At initialization phase, attractor and repoussoir are filled with randomly selected edges that are assigned an attractive/repulsive value of 0.5;
- In the repoussoir case, M edges, for each tour in the pool, are selected using a tournament technique. Edges entering tournament are randomly selected in the tour, the winner being that with highest repulsive value (if an edge is not in the repoussoir list, it is given a repulsive value of 0.5). The tour is then perturbed using the selected edges that are removed from the original tour. This has the effect of breaking the tour in several connected components, some of them being reduced to a single node;
- In the attractor case, M edges, for each tour in the pool, are selected using a tournament technique. Wwe have implemented two methods:
  1. Attractor1: edges entering tournament are randomly selected in the tour, the winner being that with smallest attractive value (if an edge is not in the attractor list, it is given a attractive value of 0.5); The tour is then perturbed using the selected edges that are removed from the original tour (this method is the closest to the binary case: mutating genes less like the attractor genes).
  2. Attractor2: edges entering tournament are randomly selected in the attractor, the winner being that with highest attractive value (if an edge is not in the attractor list, it is given a attractive value of 0.5); Selected edges are then added to the original tour, any original edges adjacent to them being deleted (this is mutation by copying attractor genes, with a risk of premature convergence).
- The previous steps have the effect of breaking the tour in several connected components, some of them being reduced to a single node/city. The connected components are then glued together by pairs. This part of the algorithm has been implemented in a straightforward way, since it's not the main focus of the paper. Reconstruction is done by selecting nearest pairs of connected components and connecting them. It thus follows the general lines of the previously cited EAX algorithm, but is less refined.

Attractor2 and Repossoir methods are illustrated in Fig. 1 (there is no figure for attractor1 since it is the same method as repoussoir, with a different selection of edges to be deleted).

The *repoussoir* individual $\mathcal{R}$ is updated in the following way (see also Figure in Appendix):

1. Each repulsive value in $\mathcal{R}$ is decremented (multiplied by a α value with $\alpha < 1$). This can be viewed as an evaporation phase;
2. As in the binary case, according to their fitness value the B worst tours are selected, the number of occurences of each edge is computed and normalized (divided by $2 * B$, the maximum number of occurences): this gives their memory contribution, called β;

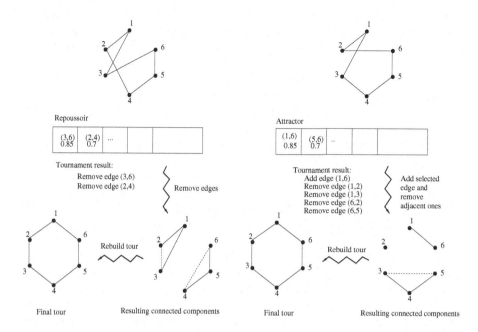

Fig. 1. Attractor2 and Repoussoir schemes for TSP.

3. Either the edge is already present in $\mathcal{R}$ or it is absent. In the first case, $(1 - \alpha) * \beta$ is added to the repulsive value of the edge in the list, in the second case, we organise a tournament between edges already in the repoussoir for the less repulsive one. If the tournament loser has a smaller repulsive value than the memory contribution β of the candidate edge, the candidate replaces it with repulsive value β.

So, to avoid the cumbersome memory size required for storing every possible edge, we are lead to remember only a small subset of those we encounter. This way allows to avoid the problem of organizing a tournament between $O(N^2)$ members. Moreover, when dynamically replacing the contents of repoussoir memory along the algorithm run, we keep on focusing on interesting edges, since bad edges from first generations are unlikely to be met again in the last stages of optimization, so we hope that they can be safely replaced. The attractor update is organized along similar lines.

4.3   Results and Conclusions for the TSP

Being perhaps the most studied problem in combinatorial optimization, the TSP has been the target of thousands of heuristics variants. It is not the objective of this paper to compete with top of the art methods. Our interest is rather in determining if repoussoir-like memory models (and notably the "shepherd and sheepdog") are able to favourably influence the search when transposed in a non binary setting. In order to answer this question, we have performed benchmarks on instances from the TSPLIB, and compared the results from four strategies:

1. the RAND method: a "classical" $(\lambda + \mu)$ evolution scheme, with no memory except the individuals themselves. It works in the same global manner as the repoussoir, but its mutation operator deletes edges chosen at random in the individual;
2. the REP method: a repoussoir only, working along the lines previously explained;
3. the SHEP1 method: a shepherd and sheepdog with repoussoir as in the REP method, and attractor1 scheme (delete edges dissimilar to attractor);
4. the SHEP2 method: a shepherd and sheepdog with repoussoir as in the REP method, and attractor2 scheme (copy edges from the attractor);

This comparison is intended to discriminate between the effects that arise from the gluing technique used in reconstructing tours, and those effects that come from the attractor/repoussoir memory model that we are interested in.

The results are presented in Tab. 2, with parameters $\mu = 10$, $\lambda = 30$, $T = 10$, $M = 10$, $B = 6$, $G = 5$. The number of function evaluations was set to 30000 for all methods. The optimal tour lengths are given on first line (italic font). Results are tour lengths averaged on 5 runs, with standard deviation between parenthesis.

Table 2. TSP results for the shepherd and sheepdog.

ALGO	TEST INSTANCES		
	lin105	**pr124**	**u159**
Optim.	*14379*	*59030*	*42080*
RAND	14578 (143)	61679 (1187)	46281 (2299)
REP	**14427** (15)	**60166** (162)	49070 (1076)
SHEP1	14667 (158)	61924 (1433)	46801 (2086)
SHEP2	14843 (230)	61991 (1892)	**46043** (1766)

As one can see there are few significant differences between strategies results, and most disparities lie between 1 standard deviation. A slight advantage may be attributed to the REP algorithm on the first two instances, both in quality and stability of results, but it falls behind the others for instance u159.

We are somewhat left to conjectures for explaining this behaviour. We are questionned about the possible inadequacy, for complex problems, of such schemes that try to target genes independantly. Nonetheless, we should acknowledge the fact that all of these strategies can provide results that are quite close to the optimum. Improving on the last one or two percents of fitness quality is undoubdtedly the hardest part of the work: it is perhaps not surprising, after all, that schemes relying on a subtle memory effect fail to take the advantage in a situation where brute exploration may be the only way. At least there is a point that seems both noticeable and explainable: the number of repoussoir mutations leading to an improvement of fitness has been noted to be much larger than the number of such successfull attractor mutations. Intuitively, we can say it is hard to obtain information on good tours, which are rare, and much easier to get usefull data about bad tours, which are encountered much more frequently. This may explain the advantage of the REP method over the SHEP and the greater standard deviation of the SHEP, keeping in mind the number of evaluations was the same for both: the SHEP could have "waste" computation time trying to get information from unreliable good tours.

The number of function evaluations was noticeably smaller than in the binary case: it may be necessary to give more computation time for these approaches to reveal themselves. Hybridizing this memory model with more classic heuristics is perhaps another way to offer perspectives for improvement.

5   Conclusion

This paper proposes an extension of the repoussoir/attractor search schemes. This extension relies on the local optimum paradigm, thus allowing an individual to change its strategy when it detects no improvement in the search. This scheme has been successfully applied to a set of academic problems in the binary search space.

We also propose an adaptation to a non binary space problem (traveling salesperson problem) of the repoussoir and attractor mecanisms. We explain the constraints that need to be dealt with, and give an implementation that keeps the semantics of the memory model. Benchmarks were run on different strategy variants. They lead to the conclusion that the kind of learning process, embedded in these repoussoir-like methods, has difficulties to give a significant advantage when dealing with the last stages of such a complex optimization problem as the TSP. Nonetheless, the repoussoir-only algorithm seems to have some potential.

References

1. Tery Jones. *Evolutionary Algorithms, Fitness, Landscapes and Search*. PhD thesis, University of Sante-Fe, Santa-Fe, New Mexico, USA, 1995.
2. D.H. Ackley. *A Connectionist Machine for Genetic Hillclimbing*. Kluwer Academic Publishers, 1987.
3. Fred Glover. Tabu search - part I. *ORSA Journal of Computing*, 1(3):190–206, 1989.
4. Shumeet Baluja. Population based incremental learning: A method for integrating genetic search based function optimization and competitive learning. Technical Report CMU-CS-94-163, Carnegie Mellon University, USA, 1994.

5. Shumeet Baluja. An empirical comparison of seven iterative and evolutionary function optimization heurustics. Technical Report CMU-CS-95-163, Carnegie Mellon University, 1995.

6. Michle Sebag and Antoine Ducoulombier. Extending population-based incremental learning to continuous search spaces. In [22], pages 418–427, 1998.

7. Michle Sebag and Marc Schoenauer. A society of hill-climbers. In [23], 1997.

8. Michle Sebag, Marc Schoenauer, and Mathieu Peyral. Revisiting the memory of evolution. Fundamenta Informaticae, 33:1–38, 1998.

9. M. Dorigo, V. Maniezzo, and A. Colorni. The ant system: Optimization by a colony of cooperating agents. IEEE Transactions on Systems, Mans, and Cybernetics, 1(26):29–41, 1996.

10. M. Dorigo and L. Gambardella. Ant colony system: A cooperative learning approach to the traveling salesman problem. IEEE Transactions on Evolutionary Computation, 1(1):53–66, 1997.

11. Jrgen Branke. Creating robust solutions by means of evolutionary algorithms. In [22], pages 119–128, 1998.

12. Cyril Fonlupt, Denis Robilliard, and Philippe Preux. A bitwise epistasis measure for binary search spaces. In [22], volume 1498 of LNCS, pages 47–56, 1998.

13. E. C. Lawler, J. K. Lenstra, and A. H. G. Rinnooy Kan, editors. The traveling Salesman Problem: A Guided Tour of Combinatorial Optimization. John Wiley and Sons, 1990. ISBN: 0-471-90413-9.

14. Guo Tao and Zbigniew Michalewicz. Inver-over operator for the TSP. In [22], pages 803–812, 1998.

15. Tim Walters. Repair and brood selection in the traveling salesman problem. In [22], pages 813–822, 1998.

16. M. Dorigo and L. Gambardella. Ant colonies for the traveling salesman problem. Technical Report TR/IRIDIA/1996-3, IRIDIA, Universit Libre de Bruxelles, Belgium, 1996.

17. Yuichi Nagata and Shigenobu Kobayashi. Edge assembly crossover: A high-power genetic algorithm for the traveling salesman problem. In [24], pages 450–457, 1997.

18. J. Watson, C. Ross, V. Eisele, J. Denton, J. Bins, C. Guerra, D. Whitley, and A. Howe. The traveling salesrep problem, edge assembly crossover, and 2-opt. In [22], pages 823–832, 1998.

19. K.D. Boese, A.B. Kahng, and S. Muddu. A new adaptive multi-start technique for combinatorial global optimizations. Operations Research Letters, 16(2):101–113, September 1994.

20. Kenneth D. Boese. Cost versus distance in the traveling salesman problem. Technical report, UCLA Computer Science Dept, Los Angeles, California, USA, 1994.

21. G. Reinelt. Tsplib — a traveling salesman problem library. ORSA J. Comput., 3:376–384, 1991.

22. Agoston E. Eiben, Thomas Bck, Marc Schoenauer, and Hans-Paul Schwefel, editors. The 5th Conference on Parallel Problem Solving from Nature, number 1498, Amsterdam, Holland, 1998. Springer-Verlag.

23. International Conference on Evolutionary Computation, Anchorage, USA, 1997.

24. Proceedings of the 7th International Conference on Genetic Algorithms, East Lansing, Michigan, USA, July 1997. Morgan Kaufmann.

A   Appendix

Initialization	$\{x_1, \ldots, x_\mu\}$ are randomly selected
	$\mathcal{A}$ (resp. $\mathcal{R}$) $= (.5, \ldots, .5)$ where $\mathcal{A}$ stands for *Attractor*, $\mathcal{R}$ for *Repoussoir*
Generation	**Pool** $= \{x_1, \ldots, x_\mu\}$
	For i $= 1$ **to** μ
	For j=1 **to** λ/μ
	Offspring = **Attractor (resp. Repoussoir)** $(x_i, \mathcal{A}(\text{resp.}\mathcal{R}))$
	Sort μ best parents + offspring
	Update Attractor (resp. Repoussoir)

Update Attractor (resp. Repoussoir)

    $y =$ average of the B best (resp. worst) individuals in Pool

    $\mathcal{A}(\text{resp.}\mathcal{R}) = (1 - \alpha) \times \mathcal{A}(\text{resp.}\mathcal{R}) + \alpha \times y$

Attractor (resp. Repoussoir)$(x, \mathcal{A}(\text{resp.}\mathcal{R}))$

    $y = x$

    For i $= 1$ **to** M where M stands for the number of bits to mutate

        $k_i =$ Tournament$(x, \mathcal{A}(\text{resp.}\mathcal{R}))$

        $y_{k_i} = 1 - y_{k_i}$

    Return y

Tournament$(x, \mathcal{A}(\text{resp.}\mathcal{R}))$

    For i $= 1$ **to** T where T parameterizes the tournament size

        Select k_i **randomly** in $\{1, \ldots, N\}$

    Return k_i such that $|x_{k_i} - \mathcal{A}_{k_i}|$ is maximum   if attractor case

    Return k_i such that $|x_{k_i} - \mathcal{R}_{k_i}|$ is minimum   if repoussoir case

Fig. 2. The Attractor (resp. Repoussoir) algorithms.

Initialization $\{x_1, \ldots, x_\mu\}$ are randomly selected
$\mathcal{A}(\text{resp.}\mathcal{R}) = (.5, \ldots, .5)$
best-so-far $= 0.0$
Hill-Climb $=$ `Repoussoir`
generation $= 0$

Generation **For** i $= 1$ **to** μ
    For j $= 1$ **to** λ/μ
        Offspring $=$ Hill-Climb(x_i)
Sort μ best parents $+$ Offspring
Update Attractor (resp. Repoussoir)
best $=$ **Select**(best *from* Pool)
If best $=$ best-so-far **then**
    generation $=$ generation $+ 1$
    If generation % $G == 0$
    Then (G defines the swap frequency)
        generation $= 0$
        Exchange(Hill-Climb)
If best $>$ best-so-far **Then**
    best-so-far $=$ best
    generation $= 0$

Exchange **If** Hill-Climb $==$ Repoussoir **Then**
    Hill-Climb $=$ Attractor
Else
    Hill-Climb $=$ Repoussoir

Fig. 3. The Shepherd and Sheepdog algorithm.

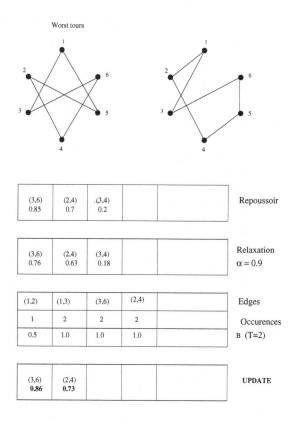

| (3,6)
0.85 | (2,4)
0.7 | .(3,4)
0.2 | | | Repoussoir |

| (3,6)
0.76 | (2,4)
0.63 | (3,4)
0.18 | | | Relaxation
$\alpha = 0.9$ |

(1,2)	(1,3)	(3,6)	(2,4)		Edges
1	2	2	2		Occurences
0.5	1.0	1.0	1.0		B (T=2)

| (3,6)
0.86 | (2,4)
0.73 | | | | **UPDATE** |

Fig. 4. Example of repoussoir memory update for TSP.

Author Index